RESIDENTIAL CONSTRUCTION CODES

COMPLETE HANDBOOK

Third Edition

Written by:

Lynn Underwood, MCP, CBO

Published by:

 CENGAGE

www.DeWALT.com/GUIDES

DEWALT Residential Construction Codes Complete Handbook, Third Edition
Lynn Underwood

SVP, Skills Product Management:
Jonathan Lau

Product Director: Matthew Seeley

Senior Product Manager:
Vanessa Myers

Executive Director of Development:
Marah Bellegarde

Senior Content Development Manager:
Larry Main

Content Developer: Jenn Alverson

Vice President, Strategic Marketing
Services: Jennifer Ann Baker

Marketing Manager: Scott Chrysler

Project Manager, Production Vendor
Management: James Zayicek

Content Project Management:
Lumina Datamatics, Inc.

Art Designer: Angela Sheehan

For product information and technology assistance, contact us at
Cengage Customer & Sales Support, 1-800-354-9706

For permission to use material from this text or product,
submit all requests online at **www. cengage.com/permissions.**
Further permissions questions can be e-mailed to
permissionrequest@cengage.com.

Library of Congress Control Number: 2018930195

ISBN: 978-1-3372-7141-7

Cengage
20 Channel Center Street
Boston, MA 02210
USA

Cengage is a leading provider of customized learning solutions with employees residing in nearly 40 different countries and sales in more than 125 countries around the world. Find your local representative at **www.cengage.com.**

Cengage products are represented in Canada by Nelson Education, Ltd.

To learn more about Cengage platforms and services, visit **www.cengage.com.**

To register or access your online learning solution or purchase materials for your course, visit **www.cengagebrain.com.**

NOTICE TO THE READER

Publisher and DEWALT® do not warrant or guarantee any of the products described herein or perform any independent analysis in connection with any of the product information contained herein. Publisher and DEWALT® do not assume, and expressly disclaim, any obligation to obtain and include information other than that provided to it by the manufacturer. The reader is expressly warned to consider and adopt all safety precautions that might be indicated by the activities described herein and to avoid all potential hazards. By following the instructions contained herein, the reader willingly assumes all risks in connection with such instructions. The publisher and DEWALT® make no representations or warranties of any kind, including but not limited to, the warranties of fitness for particular purpose or merchantability, nor are any such representations implied with respect to the material set forth herein, and the publisher and DEWALT® take no responsibility with respect to such material. The publisher and DEWALT® shall not be liable for any special, consequential, or exemplary damages resulting, in whole or part, from the readers' use of, or reliance upon, this material.

DEWALT® and GUARANTEED TOUGH are registered trademarks of the DEWALT® Industrial Tool Co., used under license. All rights reserved. Copyright DEWALT 2018. The yellow and black color scheme is a trademark for DEWALT® Power Tools and Accessories. Trademark Licensee: Cengage Learning, Executive Woods, 5 Maxwell Drive, Clifton Park, NY 12065, Tel.: 800-354-9706, www.ConstructionEdge.cengage.com. A licensee of DEWALT® Industrial Tools.

Printed in the United States of America
Print Number: 04   Print Year: 2019

CONTENTS

SECTION 3 **PLUMBING**   *201*

SECTION 4 **ELECTRICAL**   *271*

FOREWORD

For a contractor, builder, or owner, the most complex part of the construction process is navigating the building codes that regulate safety. Those in the building safety profession know that the codes are not that complicated once you have an understanding of them—but a builder hardly has time to learn all of the nuances of the building code while carrying out his job of building a home.

This book helps you learn and understand the most significant parts of the code that will affect your project. In this book, about 90% of all code requirements that you will encounter are depicted and explained. Imagine having an experienced building inspector right next to you, helping to guide you through the process. This book will give you the information to pass one of the dozen or so inspections the first time once your project is ready for inspection. That matters because in construction, time is money.

The book is arranged in a logical format according to the code itself. Each code provision is called out by section number to help you communicate with your inspector. Full-color pictures clarify the code requirements. The content is based on the very newest code, the 2018 edition of the *International Residential Code® (IRC)*.

A caution and reminder: many states or localities amend or alter the adopted version of the IRC. Be sure to ask your building department for a list of changes to compare against the IRC and enlist the help of your inspector to understand the changes and any other issues.

A NOTE TO THE READER

The *DEWALT Residential Construction Codes Complete Handbook* is not a formal code interpretation and is not intended to replace the code. You will need a copy of the 2018 *International Residential Code®* to check all references given in this book.

This book is designed to help you understand some, but not all, phases of residential construction. The systems illustrated in this book are not the only way to install or construct the systems shown. Local codes sometimes have exceptions or regulations that are enforced along with the 2018 *International Residential Code®*.

Anyone working on a residential construction project should contact their local building inspection department, office of planning and zoning, and/or department of permits in order to learn which codes are being used and how they will affect their project.

ABOUT THE INTERNATIONAL CODE COUNCIL

The International Code Council is a member-focused association dedicated to helping the building safety community and construction industry provide safe, sustainable, and affordable construction through the development of codes and standards used in the design, build, and compliance process. Most U.S.

communities and many global markets choose the International Codes. ICC Evaluation Service (ICC-ES), a subsidiary of the International Code Council, has been the industry leader in performing technical evaluations for code compliance, fostering safe and sustainable design and construction.

Headquarters:
500 New Jersey Avenue, NW, 6th Floor
Washington, DC 20001-2070

Regional Offices:
Birmingham, AL; Chicago, IL; Los Angeles, CA

1-888-422-7233
www.iccsafe.org

ABOUT THE AUTHOR

Lynn Underwood is a building code official in Washington DC, the Nation's capital. He has worked in the building safety profession and in building code development for 35 years. A Master Code Professional and Certified Building Official, he is fully certified in all aspects of construction by the International Code Council and has sat on four national code development committees. Over the past 18 years, he has participated in developing significant changes in the codes and regularly testifies at ICC code hearings for or against proposed code changes. He was elected to a 3-year term on the Board of Directors with the International Code Council. *He was recently elected President of ICC Region VII Chapter.* A past-president of the Virginia Building and Code Official's Association, Lynn has written five books on construction and inspection. The most recent is entitled *DEWALT® IRC Code Reference*. Other books recently published are *IRC Compliance for Builders and Inspectors*; *Your Green Home*; *Homebuilding, Debt-Free*; and *Common Code Problems* (first and second editions). He twice participated in the rewriting of the Building Department Administration book published by the ICC. In addition, he was a contributing author for *Healthy and Safe Homes*, published by the American Public Health Association in 2010. He also writes regularly for several national magazines, including *Building Safety Journal*, *Mother Earth News* and *Fine Homebuilding*.

A graduate engineer, Lynn is a former home builder and a Class A licensed contractor in Virginia. He teaches construction and building code classes at the Virginia Building Code Academy (Advanced Instructor), Eastern Shore Community College, and Tidewater Community College, and has developed extensive educational and training curricula for construction and inspection programs.

In 2005, Lynn led a team of inspectors to El Salvador to inspect restoration work performed by USAID. In 2007 and 2008, he twice traveled to Kabul, Afghanistan, as a member of the ICC Expert Consulting Team to help develop a building safety platform for the nation's rebuilding effort. In 2010, he was asked by the nation of Libya to present details of the International Energy Code and Green Building. He worked for the Emirate of Abu Dhabi as an Expert Building Consultant.

Lynn's hobbies include construction, fine woodworking, and photography, reading, writing, and amateur astronomy. There is an asteroid named in his honor (Underwood 15294-1991 VDB) for his volunteer construction work on an observatory in Tucson, Arizona. Lynn is a decorated, former Marine and veteran of the Vietnam War.

SECTION 1

CONSTRUCTION

PARTS OF THE INTERNATIONAL RESIDENTIAL CODE

Part I: Administrative	Includes Chapter 1, "Administration and Enforcement": covering administrative provisions such as department organization, permits, construction documents, fees, inspections, certificate of occupancy, violations, board of appeals, and stop work orders.
Part II: Definitions	Includes Chapter 2: definitions of terms used in this code.
Part III: Building Planning and Construction	Includes Chapters 3–10: covers such topics as building planning, foundations, floors, walls, roof, roof covering, chimneys, and fireplaces.
Part IV: Energy	This chapter refers to energy efficiency requirements for one and two family dwellings and is extracted from the *International Energy Conservation Code®*. It includes prescriptive provisions for the building thermal envelope, energy-using systems, and lighting systems. It also provides for compliance with performance measures.
Part V: Mechanical	Includes Chapters 12–23: heating and cooling equipment, exhaust systems, duct systems, combustion air, chimneys and vents, special fuel-burning equipment, boilers, water heaters, hydronic piping, and special piping and storage systems.
Part VI: Fuel Gas	Includes Chapter 24: structural safety, appliance location, combustion, ventilation and dilution air, equipment installation, clearance reduction, electrical connection and bonding, pipe size, materials and system installation, piping bends, inspection, piping support, vents, and specific considerations for fuel-fired appliances.
Part VII: Plumbing	Includes Chapters 25–33: plumbing fixtures, water heaters, water supply, sanitary drainage vents, traps, and storm drainage.
Part VIII: Electrical	Includes Chapters 34–43: definitions, services, branch circuit and feeder requirements, wiring methods, power and lighting distribution, devices, luminaires, appliance installation, and swimming pools.
Part IX: Referenced Standards	Includes Chapter 44: hundreds of standards from over 40 standard-developing organization such as ASTM, ASHRAE, ASCE, NSF, and UL.
Part X: Appendices	Includes Appendices A–U: alternate methodology for sizing and capacity of gas piping, venting systems, exit terminals of mechanical systems, procedure for safety inspection of existing appliance installation, manufactured housing used as a dwelling, radon control methods, swimming pools and hot tubs, private sewage disposal, existing buildings, sound transmission, permit fees, home day care, venting methods, gray water recycling systems, cross-references between the IRC and the NEC for electrical provisions, light-clay straw construction, straw bale construction, worst-case testing of atmospheric venting, and solar-ready provisions for single family dwellings.

ADMINISTRATIVE CODE PROVISIONS

Code	Title	Description
R101.3	Intent	The purpose is to provide minimum requirements to safeguard public safety and health.
R102.2	Other Laws	The provisions of this code shall not nullify other laws.
R104	Powers and Duties of the Building Official	The building official is the senior officer charged with enforcing this code. As such they have the authority to *interpret* the code and adopt policies and procedures to clarify the code. They do so with other staff: plans examiners and inspectors. They issue permits and conduct inspections.
R104.6	Right of Entry	Building officials and their staff have the right to enter property to verify a code violation, provided that if such structure or premises be occupied that credentials be presented to the occupant and entry requested.
R104.10	Modifications	Building officials may grant a modification to any provision of this code. They are charged with demonstrating that the modification is in compliance with the intent and purpose of this code.
R105	Permits, Generally	Most construction work requires a permit. This includes general construction, enlargement, alteration, repair, moving, demolishing a structure, or changing the occupancy classification of a building. This includes electrical, mechanical, and plumbing work. There are several exceptions to this general rule:
R105.2	Exempt from Permit	• One-story, detached, accessory structures not over 200 ft² • Fences not over 7' in height • Retaining walls not over 4' in height • Water tanks supported on grade not over 5000 gallons capacity • Sidewalks and driveways • Painting, papering, tiling, carpeting, cabinets, and similar trim work • Prefabricated swimming pools less than 24" deep • Certain window awnings supported by exterior wall • Decks not over 200 ft² and not more than 30" above grade (not attached to dwelling and not serving the required egress door)

You Should Know

- The IRC is a model code that can be amended by an authority having jurisdiction such as a local city, county, or state agency. Local code is the only law that matters.

- Building codes are one of numerous rules that regulate construction and include zoning, historic preservation, hillside development, environmental protection, and aesthetic as well as private sector restrictive covenants that protect a neighborhood's quality.

- States, counties, and cities adopt model codes with amendments. Certain amendments could affect exemptions from a permit.

- All of the prescriptive specifications refer to conventional materials accepted by the code. The building official is granted the authority to accept an alternative type of material even if it is not within the traditional list of materials regulated by the code. This or similar language has been in the model codes for almost a century. The purpose is to allow the building official the authority to accept new materials that may be invented in the intervening time between editions of the code. Codes almost always lag behind modern technology and innovative advancements. But how can anyone be sure that the material is safe? And since safety is paramount in the codes, this seems an odd exception in the code. It will be up to you, the contractor, developer, or owner to prove to the building official that your proposed material or manner of building is *safe*. You can do that any one of several ways, as long as they are acceptable to the building official. The most common practice is to have a registered design professional (architect or engineer) perform an analysis and substantiate the design in the form of a drawn plan that may include structural calculations. Engineering and architecture includes the science of construction. The elements of structural design are much more elaborate than the prescriptive aspects of the *International Residential Code*. In addition, there may be an independent evaluation based on testing available already. This testing may have engineering analysis as part of the evaluation. That testing may satisfy the conditions referenced in the code.

- There is another provision of the code you may find useful. The term *modification* means to alter, change, amend, adjust, or adapt. In the building code it means to alter, change, amend, adjust, or adapt the code itself! It allows the building official to *modify* any provision of the code whenever there are *practical difficulties* in applying the code. The building code finds itself in a constant state of playing "catch up" with new technologies. Manufacturers make these technologies and promote them at home shows. The contractor or owner-builder tests them and they eventually are considered to be mainstream. They are then adopted in the code and conditions are established for their use. That process takes years if not decades. Because of that, deviations from the code may still be approved as long as the contractor or owner demonstrates that the intent of any particular code is met by the optional method (or material) they want to use.

CONSTRUCTION

ADMINISTRATIVE CODE PROVISIONS

Code	Title	Description
R105.3	Application for Permit	To receive a permit, you must submit a written application. The application must have the following basic information: • The work to be done • The land on which the work is to be done • The use and occupancy for the proposed work • Construction documents • The valuation of the proposed work • Signature of the applicant or applicant's authorized representative • Other data required by the building official
R105.3.2, R105.4, R105.5, and R105.6	Time Limit, Expiration, Suspension or Revocation of Permit	All permits expire if no work is started within 180 days after issuance. All permits expire if work is suspended or abandoned for a period of 180 days after work has commenced. Extensions may be granted by the building official, but for periods of 180 days each. A permit may be revoked if issued in error or on the basis of incorrect information in violation of this code.
R105.7	Placement of Permit	The permit must be kept on the site of work until the completion of the project.
R105.8	Responsibility	Everyone who participates in work under this authority of this permit must comply with this code.
R110	Inspections	• Foundation inspection required before placement of concrete. This inspection verifies size and shape of foundation and any reinforcement. • Plumbing, mechanical, or electrical systems inspection after rough-in and prior to covering. This inspection consists of verifying that all utility installations are proper before appliances are set. • Floodplain inspection required in flood hazard areas. This verifies that the finish floor is above the flood elevation height. This is normally done by a registered design professional. • Frame and masonry inspection required after assembled and prior to covering. This inspection verifies that all materials are the proper size and that all connections are proper. It also includes verifying that draft-stopping, fireblocking, and structural headers, beams, columns, and bracing are installed. • Other inspections may be required by the local building official. These might include drywall installation inspection or shear wall inspection. • Final inspection is required when all work is complete. This inspection verifies that all systems are complete and operating properly. • Certificate of occupancy issued prior to occupancy.

You Should Know

- The applicant for a permit may be the owner or contractor or anyone responsible for the work being performed.
- Most administrative provisions are changed by local jurisdictions, so be sure and check before beginning work.

CONSTRUCTION DOCUMENTS

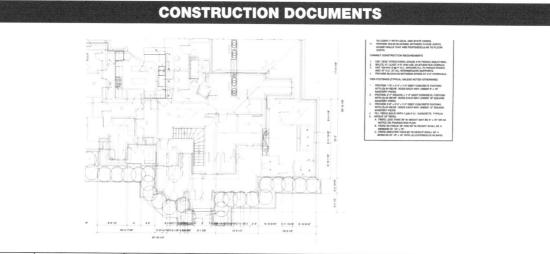

Code	Title	Description
R106.1	Construction Documents	Two (or more) sets of documents must be submitted for review with the permit application. These documents must be completed by a registered design professional (architect or professional engineer) if required by the jurisdiction. Plans may be waived by the building official.
R106.1.1	Information on Construction Documents	Plans must be of sufficient clarity to indicate the location, nature, and extent of the work. Electronic media are permitted when approved by the building official. Braced wall lines may be required to be depicted on plans.
R106.1.4	Flood Hazard Areas	If construction is within a flood hazard area, certain additional requirements must be met by the construction documents. These include: • Delineation of flood hazard areas and design flood elevation. • The elevation of the proposed lowest floor, including the basement. • The elevation of the bottom of the lowest horizontal structural member in a coastal high hazard (V) zone. • If design flood elevations are not available on the community's Flood Insurance Rate Map (FIRM), other flood elevation design and floodway data may be used.
R106.2	Site Plan	A site plan is required to be part of the construction documents. The size and location of all new construction and existing structures must be identified on the plans.
R106.3	Examination	A review of plans is made by the building official. Upon approval, the documents shall be marked: "Reviewed for Code Compliance."

You Should Know

- There are jurisdictional requirements that establish who may develop construction drawings. This may be an architect or professional engineer. In some cases, a draftsman, contractor, or the owner may draw the plan.
- When the plans are reviewed and approved for compliance with this code, a permit may be issued. Any changes in this plan must be reviewed by the building official prior to installation.
- Normally two or more sets of plans must be submitted. One is kept and one is issued to the builder. This set of plans is expected to be on-site during inspections.
- Engineered truss drawings must be submitted and approved by the building official prior to installation.

SITE PLAN

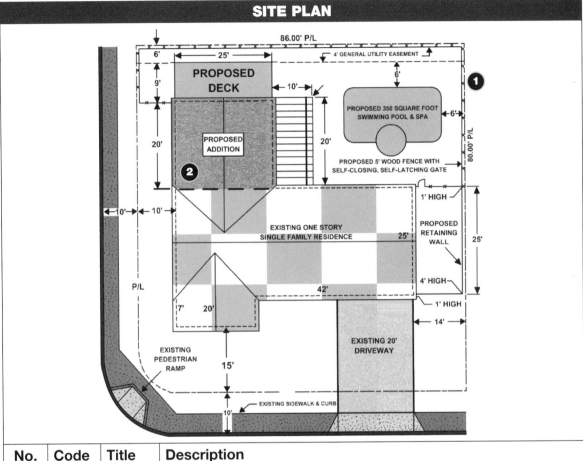

No.	Code	Title	Description
❶		Site Plan	Construction documents must include a site plan.
❷	R106.2	Site Plan	The site plan must include: • Size and location of new construction and existing construction • Distance from construction to lot lines • For demolition, construction to be demolished and remaining structures
		Site Plan	In the case of demolition, the site plan will show construction to be demolished.

You Should Know

- Jurisdiction requirements may include a variety of references.
- In any case, the plans must be drawn to scale or dimensioned.
- These may include concrete driveways, rights-of-way, including sidewalks and streets, decks, flood elevation height, and topography of the lot.
- The building official is authorized to waive or modify any requirement for a site plan where the permit application is for alteration or repair or when otherwise warranted.
- Engineered truss drawings must be submitted and approved by the building code official prior to installation.

WHO CAN DESIGN?

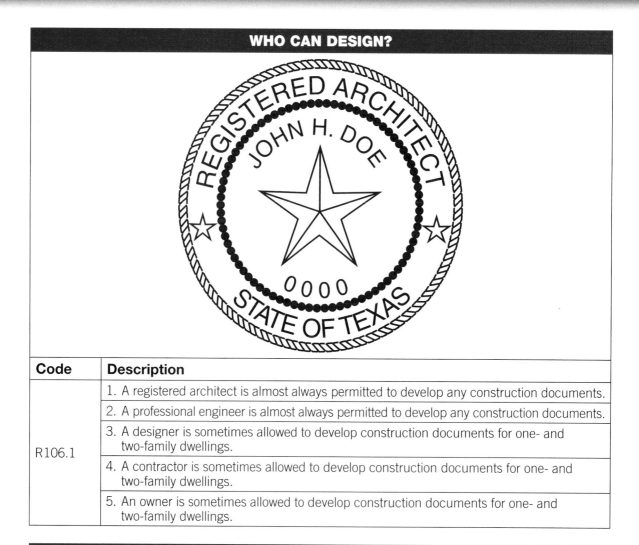

Code	Description
R106.1	1. A registered architect is almost always permitted to develop any construction documents.
	2. A professional engineer is almost always permitted to develop any construction documents.
	3. A designer is sometimes allowed to develop construction documents for one- and two-family dwellings.
	4. A contractor is sometimes allowed to develop construction documents for one- and two-family dwellings.
	5. An owner is sometimes allowed to develop construction documents for one- and two-family dwellings.

You Should Know

- Each state will have laws regulating who may create construction drawings—even for a single-family home.
- Sometimes jurisdictions will add a condition to that regulation.
- A building department may establish a plan-quality policy that restricts who may submit a plan for review.
- So, be sure and ask about this before starting construction.

PERMIT APPLICATION AND PLAN REVIEW

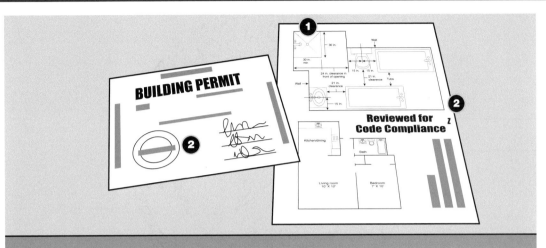

No.	Code	Description
❶	R103.3	A permit technician will be the first face a contractor sees. The technician will ask for certain information, including a completed and signed permit as well as the construction documents. The technician may also screen for other jurisdictional requirements, including zoning compliance, flood elevation height, and building setbacks from property line.
		The next step is a review of the construction documents by the plans examiner. This technical professional is often an engineer or architect who will review the plan for compliance for requirements in the IRC. This includes a review of the site plan, foundation, floor plan, stair detail, exits and guards, electrical, plumbing, and mechanical installations, structural framing (floor, wall, and roof), elevation views, and a cross section as well as details of connections.
❷	R106.3	The plan may have deficiencies and will be returned to the contractor for corrections. The deficiencies may be either missing information or incorrect details or language that is contrary to the code itself. The contractor must correct the deficiency and resubmit the plan for a rereview.
		At this point, with every correction made, the plan will be marked "Reviewed for Code Compliance" and a permit will be issued for the proposed work.

You Should Know

- Work associated with the permit must commence within 180 days of the permit issuance. At that point, the permit will expire. You may request a renewal of a permit or apply for a new permit.
- The building official is authorized to waive any construction documents not required to be prepared by a registered design professional if the work is minor in nature and review of such documents is not necessary to obtain compliance with the code.

INSPECTIONS

CONSTRUCTION

INSPECTIONS *(cont.)*

No.	Code	Description
1	R109.1.1	Inspection of the foundation shall be made after piles or piers are set or trenches or basement areas are excavated and any required forms erected and any required reinforcing steel is in place and supported prior to the placing of concrete. The foundation inspection shall include excavations for thickened slabs intended for the support of bearing walls, partitions, structural supports, or equipment and special requirements for wood foundations.
	R109.1.2	Rough inspection of plumbing, mechanical, gas, and electrical systems shall be made prior to covering or concealment, before fixtures or appliances are set or installed, and prior to framing inspection.
2	R109.1.4	Inspection of framing and masonry construction shall be made after the roof, masonry, all framing, fire-stopping, draft-stopping, and bracing are in place and after the plumbing, mechanical, and electrical rough inspections are approved.
3	R109.1.6	Final inspection shall be made after the permitted work is complete and prior to occupancy. Flood hazard areas require documentation of elevation.
	R109.1.6.1	Buildings in flood hazard areas must provide documentation of required elevations.

You Should Know

- The code calls out a few critical inspections, including those referenced. In addition, the code requires fire-resistance-rating inspection. A floodplain inspection is required if you're within a flood hazard zone. However, a jurisdiction may add even more inspections, including a drywall fastening inspection. Be sure to ask.

- Keep the approved plans, permit, and inspection records handy for an inspector. An inspector may deny the inspections if construction documents or records are not available.

- After all inspections have been approved, the building official shall issue a certificate of occupancy. The building official may issue a temporary certificate of occupancy before the completion of all of the work. The building official shall set a time period as to how long the temporary certificate of occupancy is valid.

STRUCTURAL DESIGN CRITERIA

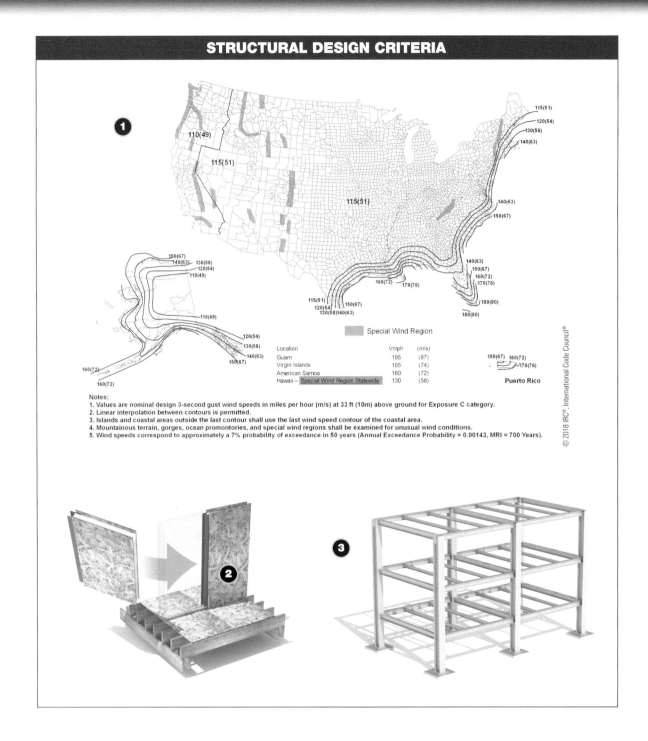

Location	Vmph	(m/s)
Guam	195	(87)
Virgin Islands	165	(74)
American Samoa	160	(72)
Hawaii – Special Wind Region Statewide	130	(58)

Special Wind Region

Puerto Rico

Notes:
1. Values are nominal design 3-second gust wind speeds in miles per hour (m/s) at 33 ft (10m) above ground for Exposure C category.
2. Linear interpolation between contours is permitted.
3. Islands and coastal areas outside the last contour shall use the last wind speed contour of the coastal area.
4. Mountainous terrain, gorges, ocean promontories, and special wind regions shall be examined for unusual wind conditions.
5. Wind speeds correspond to approximately a 7% probability of exceedance in 50 years (Annual Exceedance Probability = 0.00143, MRI = 700 Years).

© 2018 IRC®, International Code Council®

STRUCTURAL DESIGN CRITERIA (cont.)

© pics721/Shutterstock.com.

No.	Code	Description
	R301.1	Buildings constructed shall provide a complete load path from the point of origin to the foundation.
❶	R301.2.1.1	In high wind areas where design for wind is required, use AWC *Wood Frame Construction Manual for One- and Two-Family Dwellings* (WFCM) or *ICC Standard for Residential Construction in High Wind Regions*, ICC–600; or *Minimum Design Loads for Buildings and Other Structures*, ASCE-7; or *American Iron and Steel Institute (AISI) Standard for Cold-Formed Framing* (AISC S230).
		For concrete construction, wind limitations are determined in Sections R404 and R608.
❷		For structural insulated panels, wind limitations are determined in Section R610.
❸	R301.1.3	Engineered design when building of otherwise conventional construction contains structural elements exceeding the limits of this code.
❹	R301.2.1.1.1	Sun rooms must comply with AAMA/NPEA/NSA 2100 standard. There are five categories of sun room types, one of which you must select and specify on plans. This will determine code requirements elsewhere in the IRC.

You Should Know

- Wind, seismic, gravity, and flooding forces all contribute to failures in a building's structural system. Design criteria set out minimum force levels for a building based on climatic or geographic conditions.
- Engineered design may use any materials or method of construction as an alternative material. The design or method of construction shall be approved where the building official finds that the proposed design is satisfactory and complies with the intent of the provisions of this code, and that the material, method, or work offered is, for the purpose intended, at least the equivalent of that prescribed in this code.
- In addition, individual portions of the IBC may be used as an alternative to any prescriptive aspects of the IRC.
- The engineered design only need to demonstrate those components that are outside of the scope of non-conventional framing. The balance of the structure may be built under the conventional framing methods of the IRC.

LIVE AND DEAD LOADS

TABLE R301.5 MINIMUM UNIFORMLY DISTRIBUTED LIVE LOADS (IN POUNDS PER SQUARE FOOT) ①

Use	Live Load
Uninhabitable attics without storage[b]	10
Uninhabitable attics with limited storage[b,g]	20
Habitable attics and attics served with fixed stairs	30
Balconies (exterior) and decks[e]	40
Fire escapes	40
Guardrails and handrails[d]	200[h]
Guardrail in-fill components[f]	50[h]
Passenger vehicle garages[a]	50[a]
Rooms other than sleeping room	40
Sleeping rooms	30
Stairs	40[c]

© 2018 IRC®, International Code Council®

For SI: 1 pound per square foot = 0.0479 kPa, 1 square inch = 645 mm², 1 pound = 4.45 N.

a. Elevated garage floors shall be capable of supporting a 2,000-pound load applied over a 20-square-inch area.
b. Uninhabitable attics without storage are those where the maximum clear height between joists and rafters is less than 42 inches, or where there are not two or more adjacent trusses with web configurations capable of accommodating an assumed rectangle 42 inches high by 24 inches in width, or greater, within the plane of the trusses. This live load need not be assumed to act concurrently with any other live load requirements.
c. Individual stair treads shall be designed for the uniformly distributed live load or a 300-pound concentrated load acting over an area of 4 square inches, whichever produces the greater stresses.
d. A single concentrated load applied in any direction at any point along the top.
e. See Section R502.2.2 for decks attached to exterior walls.
f. Guard in-fill components (all those except the handrail), balusters, and panel fillers shall be designed to withstand a horizontally applied normal load of 50 pounds on an area equal to 1 square foot. This load need not be assumed to act concurrently with any other live load requirement.
g. Uninhabitable attics with limited storage are those where the maximum clear height between joists and rafters is 42 inches or greater, or where there are two or more adjacent trusses with web configurations capable of accommodating an assumed rectangle 42 inches in height by 24 inches in width, or greater, within the plane of the trusses.
   The live load need only be applied to those portions of the joists or truss bottom chords where all of the following conditions are met:
   1. The attic area is accessible from an opening not less than 20 inches in width by 30 inches in length that is located where the clear height in the attic is a minimum of 30 inches.
   2. The slopes of the joists or truss bottom chords are no greater than 2 inches vertical to 12 units horizontal.
   3. Required insulation depth is less than the joist or truss bottom chord member depth.
   The remaining portions of the joists or truss bottom chords shall be designed for a uniformly distributed concurrent live load of not less than 10 lb/ft².
h. Glazing used in handrail assemblies and guards shall be designed with a safety factor of 4. The safety factor shall be applied to each of the concentrated loads applied to the top of the rail, and to the load on the infill components. These loads shall be determined independent of one another, and loads are assumed not to occur with any other live load.

LIVE AND DEAD LOADS *(cont.)*

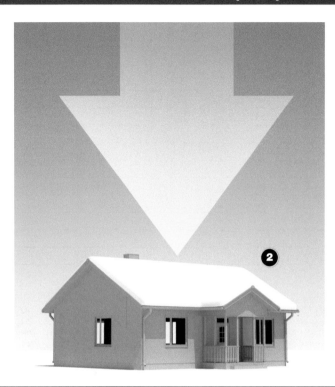

2 TABLE R301.7 ALLOWABLE DEFLECTION OF STRUCTURAL MEMBERS[b, c]

Structural Member	Allowable Deflection
Rafters having slopes greater than 3:12 with finished ceiling not attached to rafters	$L/180$
Interior walls and partitions	$H/180$
Floors	$L/360$
Ceilings with brittle finishes (including plaster and stucco)	$L/360$
Ceilings with flexible finishes (including gypsum board)	$L/240$
All other structural members	$L/240$
Exterior walls—wind loads[a] with plaster or stucco finish	$H/360$
Exterior walls—wind loads[a] with other brittle finishes	$H/240$
Exterior walls—wind loads[a] with flexible finishes	$H/120$[d]
Lintels supporting masonry veneer walls[e]	$L/600$

© 2018 IRC®, International Code Council®

Note: L = span length, H = span height.

a. For the purpose of determining deflection limits herein, the wind load shall be permitted to be taken as 0.7 times the component and cladding (ASD) loads obtained from Table R301.2(2).
b. For cantilever members, L shall be taken as twice the length of the cantilever.
c. For aluminum structural members or panels used in roofs or walls of sunroom additions or patio covers, not supporting edge of glass or sandwich panels, the total load deflection shall not exceed $L/60$. For continuous aluminum structural members supporting edge of glass, the total load deflection shall not exceed $L/175$ for each glass lite or $L/60$ for the entire length of the member, whichever is more stringent. For sandwich panels used in roofs or walls of sunroom additions or patio covers, the total load deflection shall not exceed $L/120$.
d. Deflection for *exterior walls* with interior gypsum board finish shall be limited to an allowable deflection of $H/180$.
e. Refer to Section R703.8.2.

LIVE AND DEAD LOADS *(cont.)*

③ TABLE R301.6 MINIMUM ROOF LIVE LOADS IN POUNDS-FORCE PER SQUARE FOOT OF HORIZONTAL PROJECTION

Roof Slope	Tributary Loaded Area in Square Feet for Any Structural Member		
	0 to 200	201 to 600	Over 600
Flat or rise less than 4 inches per foot (1:3)	20	16	12
Rise 4 inches per foot (1:3) to less than 12 inches per foot (1:1)	16	14	12
Rise 12 inches per foot (1:1) and greater	12	12	12

For SI: 1 square foot = 0.0929 m², 1 pound per square foot = 0.0479 kPa, 1 inch per foot = 83.3 mm/m.

© 2018 IRC®, International Code Council®

No.	Code	Description
①	R301.4 and R301.5	Live and dead loads: Table R301.6 depicts design live loads for various conditions around a home. These are used in various tables for sizing structural elements such as rafters, joists, and wall framing. Generally, this table is limited to floors, walking surfaces, or exit systems.
	R301.2.3	Prescriptive design for snow loads limited to 70 psf or less.
②	R301.7	Deflection of a structural member is limited based on loading condition and location.
③	R301.6	Roof live loads are based on tributary loaded area and roof slope. This load is used in determining the size of structural elements such as rafters in various tables in this code (unless snow load controls).

You Should Know

- All prescriptive design in the IRC is based on loading forces such as live and dead loads. These vary according to the use or location of the structural member.
- Deflection of a structural member is limited to an allowable deflection based on the location and use of the structural member.

PROXIMITY TO PROPERTY LINE (FIRE SEPARATION DISTANCE)

TABLE R302.1(1) EXTERIOR WALLS

❶ EXTERIOR WALL ELEMENT		MINIMUM FIRE-RESISTANCE RATING	MINIMUM FIRE-SEPARATION DISTANCE
Walls	Fire-resistance rated	1 hour—tested in accordance with ASTM E 119 or UL 263 with exposure from both sides	< 5 feet
	Not fire-resistance rated	0 hours	≥ 5 feet
Projections	Not allowed	N/A	< 2 feet
	Fire-resistance rated	1 hour on the underside[a, b]	≥ 2 feet to < 5 feet
	Not fire-resistance rated	0 hours	≥ 5 feet
Openings in walls	Not allowed	N/A	< 3 feet
	25% maximum of wall area	0 hours	3 feet
	Unlimited	0 hours	5 feet
Penetrations	All	Comply with Section R302.4	< 3 feet
		None required	3 feet

For SI: 1 foot = 304.8 mm.

N/A = Not Applicable.

a. Roof eave fire-resistance rating shall be permitted to be reduced to 0 hours on the underside of the eave if fireblocking is provided from the wall top plate to the underside of the roof sheathing.

b. Roof eave fire-resistance rating shall be permitted to be reduced to 0 hours on the underside of the eave provided that gable vent openings are not installed.

© 2018 IRC®, International Code Council®

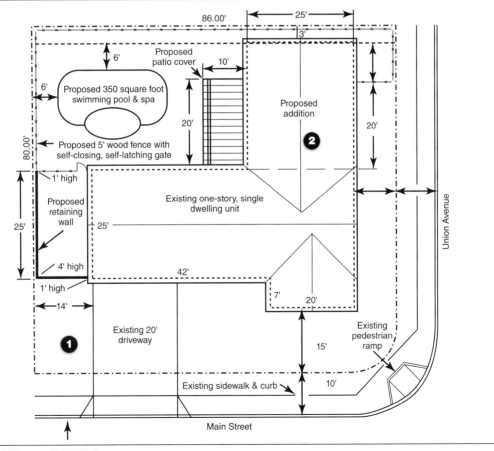

PROXIMITY TO PROPERTY LINE (FIRE SEPARATION DISTANCE) *(cont.)*

③ TABLE R302.6 DWELLING/GARAGE SEPARATION

SEPARATION	MATERIAL
From the residence and attics	Not less than ½-inch gypsum board or equivalent applied to the garage side
From all habitable rooms above the garage	Not less than ⅝-inch Type X gypsum board or equivalent
Structure(s) supporting floor/ceiling assemblies used for separation required by this section	Not less than ½-inch gypsum board or equivalent
Garages located less than 3 feet from a dwelling unit on the same lot	Not less than ½-inch gypsum board or equivalent applied to the interior side of exterior walls that are within this area

For SI: 1 inch = 25.4 mm, 1 foot = 304.8 mm.

© 2018 IRC®, International Code Council®

No.	Code	Description
❶	R302.1	Exterior walls must be at least 5′ from property line or 1 hour rated.
❷	R302.1	Windows near property line are regulated. If the wall is less than 3′, they are not permitted. If less than 5′ (and not less than 3′), they are limited to 25% of the wall area.
❸	R302.5 and R302.6	Dwelling/garage openings must be protected with 1⅜″-thick solid wood or honeycomb-core steel doors or 20 minute rated with a self-closing device.

You Should Know

- Walls perpendicular to property lines do not need to meet fire-resistance ratings.
- Dwellings with sprinklers may have a reduced setback; see Table R301.1(2). This table is not in this text, but appears in the IRC code book. It essentially reduces the setback for buildings with sprinklers from 5′ to 3′ for any nonrated walls.

HEIGHT AND STORY LIMIT

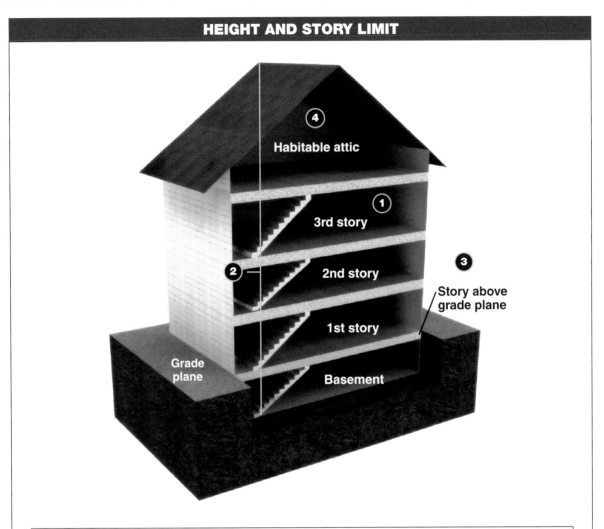

Definition of Story: That portion of a building included between the upper surface of a floor and the upper surface of the floor or roof next above.

Definition of Story Above Grade Plane: Any *story* having its finished floor surface entirely above **grade plane**, or in which the finished floor surface of the floor next above is:
1. more than 6' above grade plane or
2. more than 12' above the finished ground level at any point.

Definition of Basement: A story that is not a story above grade plane.

Definition of Habitable Attic: A finished or unfinished space within an attic.

HEIGHT AND STORY LIMIT *(cont.)*

No.	Code	Description
1	R101.2 and Chapter 2	Dwellings are limited to three stories above grade plane by the scope of the IRC. A story is defined as: That portion of a building included between the upper surface of a floor and the upper surface of the floor or roof next above.
2	R101.2 and Chapter 2	Definitions of the words *story, story above grade plane, basement,* and *habitable attic* allow for a three-story dwelling to have an additional basement and a habitable attic (5 levels) within the scope of the IRC.
3	Chapter 2	That portion of a building that is partly or completely below grade (see "story above grade").
4	Chapter 2	A habitable attic is defined as a finished or unfinished area, not considered a *story.*

You Should Know

- There are popular roof truss configurations that accommodate a full-height attic. These may be used and not be considered as a story for the purposes of maximum number of stories.
- Basement (if habitable) and habitable attic must meet the conditions for a habitable space, including heating, lighting, ventilation, ceiling height, minimum room areas, means of egress, stairs, smoke alarms, and other requirements.

FIRE PROTECTION OF FLOORS

No.	Code	Description
	R302.13	Floor assemblies even if not required to be rated elsewhere in this code shall be provided with a ½″ gypsum wallboard membrane or equivalent on the underside of the floor framing with some exceptions. Penetrations or openings for ducts, vents, electrical outlets lighting, devices, luminaires, wires, speakers, drainage, piping, and similar openings or penetrations shall be permitted. There are several other exceptions where this rule does not apply. For example, protection is not required for 2 × 10 floor joists.

CONSTRUCTION

LIGHTING, VENTILATION, AND HEATING

X × Y = min. 4% of the
Floor Area *(1.5 sq. ft. min.)*

60°F   2'

3'

No.	Code	Description
1	R303	All habitable rooms must have 8% of their floor area attributed to glazing. Natural ventilation is required through windows, doors, or other approved openings. At least 4% of floor area must be attributed to openings to the outside for this ventilation. An exception allows artificial light to substitute for this requirement.
2	R303.3	Bathrooms, water closets, compartments, and similar rooms must have at least 3 sq. ft. of glazing. Half of this must be openable (1.5 ft²). This same exception applies for artificial light and mechanical ventilation.
	R303.4	Where the air infiltration rate is less than five/air changes per hour (blower door), the dwelling shall be provided with whole-house mechanical ventilation.
3	R303.8	All exterior and interior stairways must be illuminated with an artificial light source.
4	R303.10	When winter design temperature is below 60°, heating facilities that maintain a room temperature of 68° are required at a point 2' from the wall and 3' above the floor.

You Should Know

- Requirements for HVAC systems are located in Section 2 of this guide.
- Certain rooms may enjoy the light from another room. Adjoining rooms adjacent to a room with natural illumination with half of the common wall open are considered to meet this requirement.
- Glazed areas need not be openable where the opening is not required as an emergency egress and a whole-house mechanical ventilation system is installed.

MINIMUM ROOM SIZE AND SHAPE AND CEILING HEIGHT

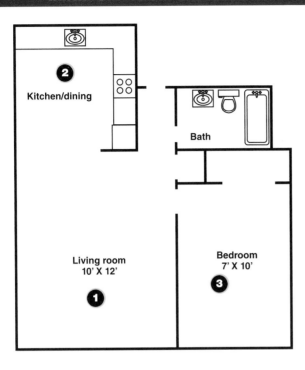

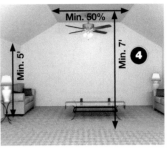

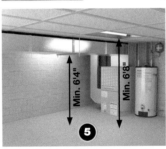

No.	Code	Description
1	R304	Habitable rooms shall have a floor area of not less than 70 ft^2.
2		Kitchens, although a habitable room, do not need to meet this criteria.
3	R304.3	Habitable rooms must not be less than 7′ in any dimension, except kitchens.
4	R304.4	Areas of rooms with a sloping ceiling measuring less than 5′ or a furred ceiling less than 7′ in height may not be considered as part of the minimum required habitable floor area.
5	R305	Generally, ceiling height must be 7′ in habitable spaces, hallways, and portions of the basement containing these uses. There are some exceptions. For instance, bathrooms, laundry rooms, and toilet rooms must only be 6′8″ tall.

You Should Know

- The ceiling height above bathroom and toilet fixtures shall be such that the fixture is capable of being used for its intended purpose.
- Required height for basements without habitable space is reduced to 6'8". Beams, girders, ducts, or other similar obstructions may project to within 6' 4" for all basements.

SANITATION

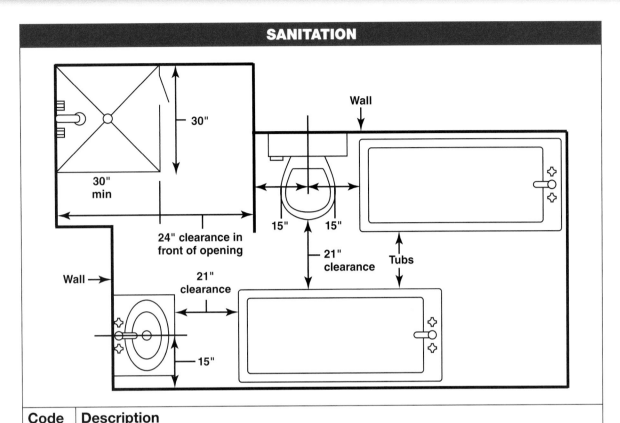

Code	Description
R306	Every dwelling unit must have a water closet, lavatory, and either a tub or a shower.
	Every dwelling must have a kitchen with a sink.
	All plumbing facilities must be connected to a sanitary sewage system or an approved private sewage disposal system (septic tank).
	All plumbing fixtures must be connected to an approved water supply. Kitchen sinks, lavatories, bathtubs, showers, bidets, laundry tubs, and washing machines must have both hot and cold water.

You Should Know

- Bathroom design is also regulated by Section R307 and Figure R307.1 in the IRC code book for access and clearances for plumbing fixtures.

SAFETY GLAZING AND HAZARDOUS LOCATIONS

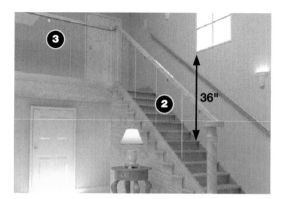

No.	Code	Description
❶		A label is required on glazing located in hazardous locations. The following are considered hazardous locations that require safety glazing:
❷	R308	**Glazing adjacent stairs and ramps.** Glazing where the bottom exposed edge of the glazing is less than 36″ (914 mm) above the plane of the adjacent walking surface of stairways, landings between flights of stairs and ramps shall be considered a hazardous location. In addition, glazing within 60″ of a horizontal arc from the bottom tread nosing.
❸		Glazing in guards and railings regardless of area or height above the walking surface.
❹		Glazing in panels of swinging and sliding doors, glazing in panels adjacent to and within a 24″ arc of a door, in a closed position where the bottom edge of the glazing is less than 60″ above the floor. Glazing in all swinging, sliding, or bifold doors shall be considered hazardous condition.
❺	R308.4.5	Glazing for walls facing hot tubs, swimming pools, whirlpools, saunas, steam rooms, bathtubs, and showers where the bottom edge is less than 60″ above any standing or walking surface shall be considered a hazardous location. **Exception:** Glazing that is more than 60″ (1524 mm), measured horizontally and in a straight line, from the water's edge of a bathtub, hot tub, spa, swimming pool, or whirlpool, or from the edge of a shower, sauna, or steam room.

SAFETY GLAZING AND HAZARDOUS LOCATIONS *(cont.)*

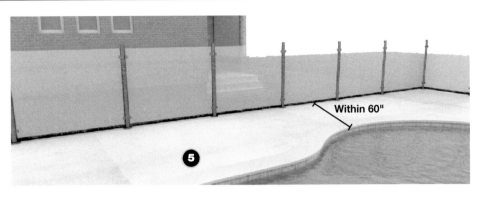

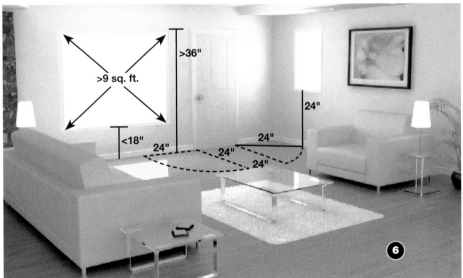

No.	Code	Description
6	R308.4.3	Glazing where the bottom exposed edge of the glazing is less than 18″ above the floor, the exposed area of the individual pane is larger than 9 sq. ft., and one or more walking surfaces are within 36″ of the glazing. See the exceptions to this general rule.
	R308.6.2	Permitted materials for skylights and sloped glazing to be used as safety glazing includes laminated glass, fully tempered glass, heat-strengthened glass, wired glass, and approved rigid plastics.

You Should Know

- There are several exceptions to the requirement for safety glazing in hazardous locations.
- Skylights installed in a roof with a slope flatter than 3:12 must be installed with a curb extending 4″ above the roof unless specified by manufacturer.
- Site-built windows are permitted but they must meet criteria in the IBC (Section 2404).

EMERGENCY ESCAPE AND RESCUE OPENINGS

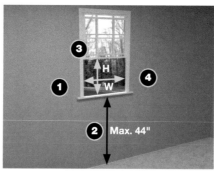

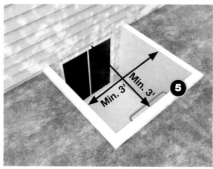

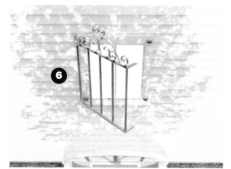

No.	Code	Description
1	R310.2.1	Openings must be 5.7 ft² for upper stories and 5.0 ft² for grade-floor openings.
2	R310.2.2	The sill of the emergency escape and rescue opening window must be no higher than 44″ from the finished floor.
3	R310.2.1	The minimum opening height must be at least 24″.
4	R310.2.1	The minimum opening width must be at least 20″.
5	R310.2.3	Window wells must be at least 9 ft² and a minimum horizontal projection and width of 36″.
6	R310.4	Bars, grills, covers, and screens are permitted to be placed over openings, enclosures, or area wells provided the net clear opening size complies with Sections R310.1.1 to R310.2.3 of the IRC. Note that Section R310.1.1 references an ASTM Standard F2090 for window opening control devices used for child fall protection. Meeting the ASTM Standard F2090 means that hardware is approved for EERO.

You Should Know

- Basements, habitable attics, and every sleeping room must have at least one emergency egress and rescue opening. The opening can be a window or a door.
- Where basements contain more than one sleeping area, each must have a separate emergency egress and rescue opening.
- Window wells must be designed for proper drainage.

EGRESS AND EXITS

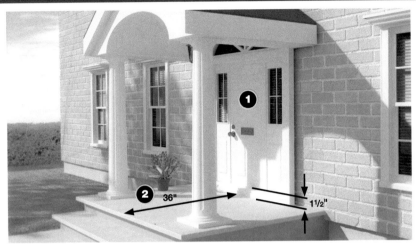

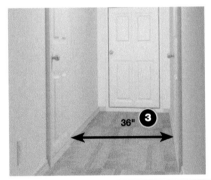

No.	Code	Description
1	R311.2	All dwellings must have at least one qualifying exit door to the outside. The door must be side-hinged and provide a net clear width of 32" between the face of the door and the stop when the door is 90° open. The minimum height must be 78" between the top of the threshold and the bottom of the stop.
2	R311.3	Floor or a landing is required on each side of the door. The width must be not less than the door served and the length must be at least 36" in the direction of travel. The landing's slope can be no more than a 2% slope or ¼ vertical units to 12 horizontal units. Floor or landing must be no lower than 1½" from the top of the threshold with one exception.
3	R311.6	Hallways must be at least 36" in width.
	R311.4	Vertical egress from habitable levels without an egress door must be by a ramp or a stair.
4	R311.2	The one required egress door shall be readily openable from the inside without the use of a key or special knowledge or effort.

You Should Know

- A landing or floor on the exterior side may be up to 7¾" below the top of the threshold as long as the door does not swing over the landing.

STAIRS, GENERALLY

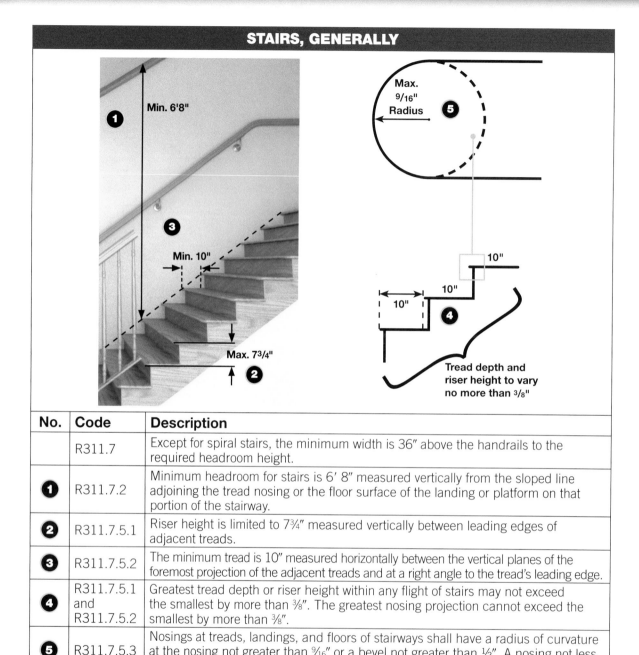

No.	Code	Description
	R311.7	Except for spiral stairs, the minimum width is 36" above the handrails to the required headroom height.
1	R311.7.2	Minimum headroom for stairs is 6' 8" measured vertically from the sloped line adjoining the tread nosing or the floor surface of the landing or platform on that portion of the stairway.
2	R311.7.5.1	Riser height is limited to 7¾" measured vertically between leading edges of adjacent treads.
3	R311.7.5.2	The minimum tread is 10" measured horizontally between the vertical planes of the foremost projection of the adjacent treads and at a right angle to the tread's leading edge.
4	R311.7.5.1 and R311.7.5.2	Greatest tread depth or riser height within any flight of stairs may not exceed the smallest by more than ⅜". The greatest nosing projection cannot exceed the smallest by more than ⅜".
5	R311.7.5.3	Nosings at treads, landings, and floors of stairways shall have a radius of curvature at the nosing not greater than ⁹⁄₁₆" or a bevel not greater than ½". A nosing not less than ¾" nor greater than 1¼" must be on stairs with solid risers.

You Should Know

- Risers must be vertical or sloped. See Section R311.7.5.1 in the IRC code book.
- The vertical rise on any flight of stairs may not exceed 151".
- Stairways shall not be less than 36" in clear width at all points above the permitted handrail height and below the required headroom height.
- Open risers on stairs are permitted provided that the openings located greater than 30" vertically to the floor do not permit the passage of a 4" diameter sphere. See Section R311.7.5.1 of the IRC.

STAIR HANDRAILS AND LIGHTING

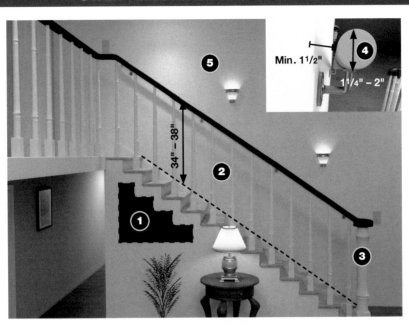

No.	Code	Description
❶	R311.7.8	Handrails must be provided at least on one side of stairs with four or more risers.
❷	R311.7.8.1	Handrail height, measured vertically from the sloped plane adjoining the tread nosing or finish surface of ramp slope, must not be less than 34″ and not more than 38″.
❸	R311.7.8.4	Handrails for stairs must be continuous for the full length of the flight. Handrail ends must be returned or end in newel posts or safety terminals. Handrails adjacent to the wall must be spaced at least 1½″ from the wall.
❹	R311.7.8.5	Grip size for handrails is regulated by the type of handrail: Type I or Type II. Type I handrails with a circular cross section must have a cross-sectional diameter between 1¼″ and 2″. Those that are not circular are limited to a perimeter between 4″ and 6¼″ and shall have a maximum cross-sectional dimension of 2½″. Type II handrails with a perimeter greater than 6¼″ must have a graspable finger recess area on both sides of the profile and specific dimensions for the recessed area.
	R311.7.8.4 and R507.3	Wood/plastic composite handrails must follow ASTM D-7032 and manufacturer's instructions when used in exterior locations.
❺	R311.7.9 and R303.7	All interior stairs must have a means to illuminate the treads and landings. Interior stairs must have an artificial light source capable of 1 foot-candle measured at the center of the treads. For exterior stairs, the light source must be near the top landing.

You Should Know

- Handrails are divided into Types I and II. The types are distinguished by perimeter size and cross-sectional shape in the IRC code book. See Section R311.7.8.5 for more details on handrails.

SPECIAL STAIRWAYS

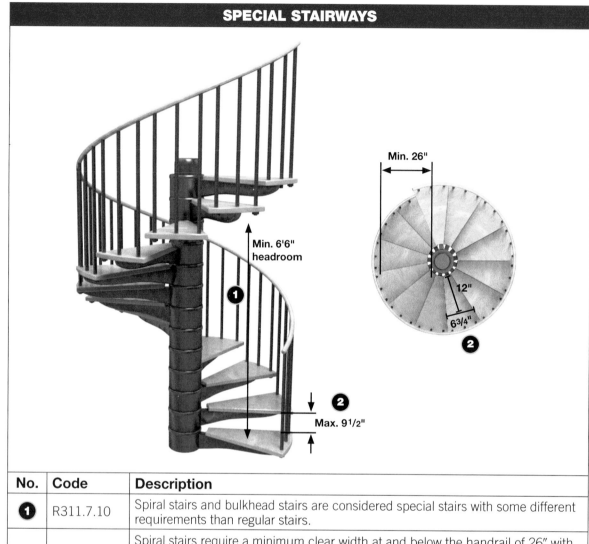

Min. 6'6" headroom

Max. 9½"

Min. 26"

12"

6¾"

No.	Code	Description
❶	R311.7.10	Spiral stairs and bulkhead stairs are considered special stairs with some different requirements than regular stairs.
❷	R311.7.10.1	Spiral stairs require a minimum clear width at and below the handrail of 26" with each tread having a minimum tread of 6¾" and minimum tread depth at 12" from the narrower side. All treads must be identical, with a maximum rise of 9½". There must be a minimum headroom of 6'6".
	R311.7.10.2	Stairways serving bulkhead enclosures, if *not* part of a means of egress, are exempt from basic requirements for width, rise and run, headroom height, etc. If the maximum height from basement finished floor to grade-level opening does not exceed 8' and the grade-level opening is covered by a bulkhead enclosure with hinged doors or other approved means.

You Should Know

- Spiral stairs built according to R311.7.10 are allowed to serve as a means of egress.

GUARDS

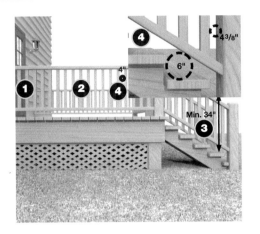
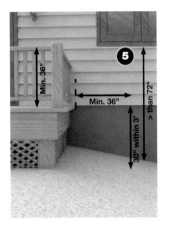

No.	Code	Description
1	R312.1	Guards shall be provided in accordance with the following:
2	R312.1.1	Guards shall be located along open-sided walking surfaces, including stairs, ramps, and landings, that are located more than 30″ measured vertically to the floor or *grade* below at any point within 36″ horizontally to the edge of the open side. Insect screening shall not be considered as a guard.
3	R312.1.2	Required guards at open-sided walking surfaces must be at least 36″ above the adjacent walking surface. Guards on the open sides of stairs must be at least 34″ above the adjacent walking surface or the line connecting the nosings.
4	R312.1.3	Required guards may not have openings between the walking surface to the required guard height that will allow a 4″ sphere to pass through. For the open side of a stair formed by the riser, the tread and bottom rail of a guard shall not allow the passage of a sphere 6″ in diameter. Guards on the open sides of stairs shall not have openings that allow the passage of a 4 ⅜″ sphere.
5	R312.2	In dwelling units where the top of the sill of an operable window opening is located less than 24″ above the finished floor and greater than 72″ above the finished grade or other surface below on the exterior of the building, the operable window shall comply with one of the following: 1. Operable windows with openings that will not allow a 40 diameter sphere to pass through the opening where the opening is in its largest opened position. 2. Operable openings that are provided with fall prevention devices that comply with ASTM F 2090. 3. Operable windows that are provided with window opening control devices that comply with ASTM F 2090.

You Should Know

- Guards prevent accidental falling from a walking surface.
- Guards are regulated for open walking surfaces, including stairs, ramps, and landings.

SPRINKLERS

No.	Code	Description
	R313.1	All dwelling units of townhouses must have automatic fire sprinklers installed.
	R313.2	An automatic residential fire sprinkler system shall be installed in one- and two-family *dwellings*.
⑥	R313.1.1 and R313.2.1	The design and installation of automatic fire sprinklers must be according to National Fire Protection Association (NFPA) 13d (Standard), Section P2904 in the IRC code book.

You Should Know

- If your jurisdiction has adopted this portion of the code, automatic fire sprinklers are required to be installed.
- The design and installation may be by either the NFPA standard for sprinklers or a section in the IRC code book for these systems. This new standard is prescriptive and follows a logical approach to the installation. It uses the domestic water supply for the home.

CONSTRUCTION

SMOKE ALARM LAYOUT

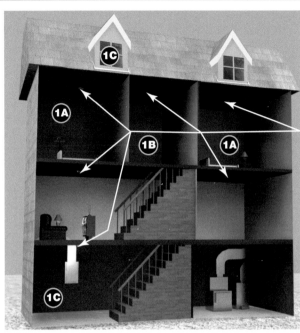

Smoke alarms are required in each bedroom, outside each separate sleeping area, and on each story including the basement

No.	Code	Description
	R314.1	Smoke alarms shall comply with NFPA 72 and listed in accordance with UL 217. Combination smoke and carbon monoxide alarms must be listed according to UL 2034.
1	R314.3	Smoke alarms must be installed in the following locations: A. In each sleeping room B. Outside each separate sleeping area in the immediate vicinity of the bedrooms C. On each additional story, including basements and habitable attics and not including crawl spaces or uninhabitable attics D. Smoke alarms shall be installed not less than 3' horizontally from the door or opening of a bathroom that contains a bathtub or shower unless this would prevent placement of a smoke alarm required by this section
	R314.2.2	When alterations, repairs, or additions occur that require a permit, the individual dwelling unit shall be equipped with smoke alarms located as required for new dwellings.
	R314.3.1	Ionization smoke alarms shall not be installed within 20' of permanently installed cooking appliance or ionization alarms with an alarm silencing switch shall not be installed less than 10' from permanently installed cooking appliance. Photoelectric smoke alarms may be reduced to 6' horizontally from permanently installed cooking appliances.

You Should Know

- When you purchase smoke alarms, be sure to review the standards of manufacture to ensure they include the required standards specified in the code.
- Many manufacturers set out additional standards for location of smoke alarms, including distance from a condition air duct or return air duct.
- Smoke alarms must be listed and labeled per UL 217 and installed according to NFPA 72.

SMOKE ALARM LAYOUT

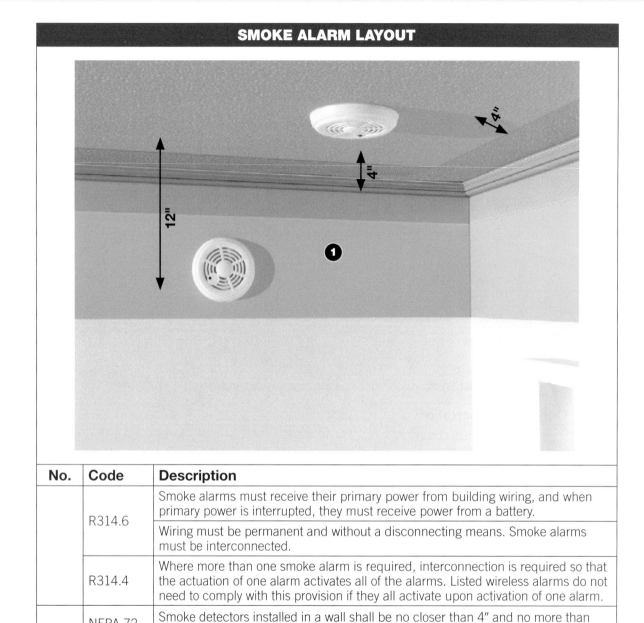

No.	Code	Description
	R314.6	Smoke alarms must receive their primary power from building wiring, and when primary power is interrupted, they must receive power from a battery.
		Wiring must be permanent and without a disconnecting means. Smoke alarms must be interconnected.
	R314.4	Where more than one smoke alarm is required, interconnection is required so that the actuation of one alarm activates all of the alarms. Listed wireless alarms do not need to comply with this provision if they all activate upon activation of one alarm.
	NFPA 72	Smoke detectors installed in a wall shall be no closer than 4″ and no more than 12″ from the ceiling.
❶	NFPA 72	When located on the ceiling, smoke detectors must be no closer than 4″ from the wall.

CARBON MONOXIDE (CO) ALARMS

Carbon monoxide detectors are now required:

Outside each separate sleeping area with dwelling units with attached garages and outside each separate sleeping area with dwelling units with fuel-fired appliances. Bedrooms or attached bathrooms with fuel burning appliances CO alarms are required inside the bedroom.

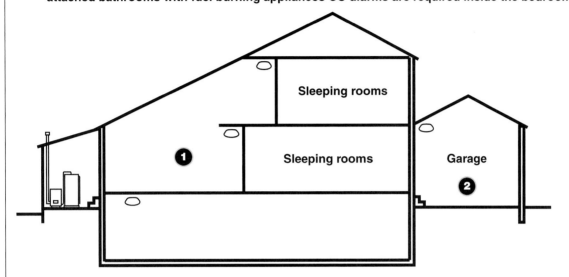

No.	Code	Description
❶	R315.1	For new construction, CO alarms are required when the dwelling unit contains a fuel-fired appliance or an attached garage with an opening that communicates to the dwelling unit.
❷	R315.2.2	Where alterations, repairs, or additions requiring a permit occur, the individual dwelling unit shall be equipped with carbon monoxide alarms located as required for new dwellings. The two exceptions for this include only work on exterior surfaces and installation of plumbing/mechanical systems.
	R315.7.1	Household CO detection systems shall comply with NFPA 720 and be listed with UL 2075.

You Should Know

- When shopping for a carbon monoxide alarm, be sure and check for the approved standard listed on the manufacturer's data sheets.

FOAM PLASTIC

No.	Code	Description
❶	R316.2	Packages and containers of foam insulation and foam plastic insulation on a jobsite must bear the label of an approved agency showing the manufacturer's name, the product listing, identification, and information about product performance.
❷	R316.3	All foam plastic or foam plastic cores must have a flame-spread index of not more than 75 and a smoke-developed index of not more than 450 according to ASTM E-84 or UL 723.
	R316.4	Foam plastic must be separated from the interior of the building by an approved thermal barrier of at least ½″ gypsum wallboard, $^{23}/_{32}$ wood structural panel, or materials that are tested in accordance with and meet the acceptance criteria of both the temperature transmission fire test and the integrity fire test of NFPA 275.
	R316.5.1	Masonry or concrete avoids the requirement for a thermal barrier.
	R316.5.2	A thermal barrier is not required for foam plastic in a roof assembly installed according to the provisions of the IRC and the roof sheathing is tongue-and-groove wood planks or $^{15}/_{32}$″-thick exposure 1, wood structural panel sheathing bonded with exterior glue.
	R316.5.3	The thermal barrier is not required in an attic or a crawl space where there is access, the space is exclusively for repairs and maintenance, and any foam plastic is protected against ignition by one of eight specific ignition barriers.

You Should Know

- Foam plastic presents a problem by off-gassing harmful chemicals during a fire. These protections allow a modest time for the occupants to escape during the early moments in a fire.

CONSTRUCTION

PRESERVATIVES

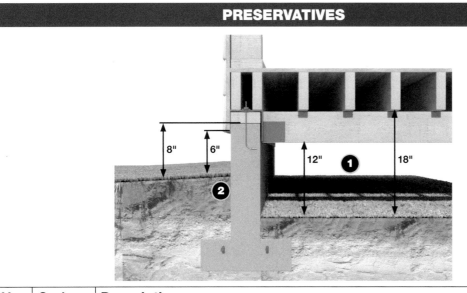

No.	Code	Description
❶	R317.1	Protection of wood and wood-based products from decay shall be provided in the following locations by the use of naturally durable wood: • Wood joists closer than 18″ and wood beams closer than 12″ from exposed ground • Wood framing members and sills and sleepers in contact with concrete or masonry less than 8″ from exposed ground • Ends of wood girders entering masonry or concrete walls with less than ½″ clearance on tops, sides, and ends • Wood siding, sheathing, and wall framing on the exterior with less than 6″ from the ground or 2″ from concrete steps, slabs, etc. • Wood structural members supporting moisture-permeable floors or roofs exposed to weather such as concrete or masonry slabs • Wood furring strips or framing members attached directly to the interior of concrete or masonry walls below grade except where an approved vapor barrier is installed per AWPA-U1.
	R317.1.1	Field-cut ends, notches, and drilled holes must be treated in accordance with the AWPA M-4 standard.
❷	R317.1.2	All wood in contact with the ground, embedded in concrete, must be approved, prescriptively treated wood suitable for ground contact.
	R317.1.4	Wood columns must be approved wood of natural resistance to decay or preservative-treated type unless elevated 1″ above concrete or 6″ above earth and covered by an approved impervious barrier.
	R317.1.5	Glu-laminated timbers exposed to weather and not properly protected must be preservative treated or manufactured with naturally durable wood.

You Should Know

- Naturally durable or preservative-treated wood is still required where local experience demonstrates a need.
- Fasteners and connectors in contact with preservative-treated wood must be hot dipped, zinc-coated, galvanized steel, stainless steel, silicon bronze, or copper.

TERMITES

No.	Code	Description
①	R318.1	In areas subject to damage from termites, one of the following methods shall be used to protect a building (also see Figure R301.2[6] in the IRC code book): • Chemical termiticide treatment • Termite baiting systems • Pressure-preservative treated wood • Naturally durable termite-resistant wood • Physical barriers • Cold-formed steel framing
	R318.4	Foam plastic shall not be installed in areas where the probability of infestation is very likely.

ADDRESS NUMBERS

No.	Code	Description
❶	R319.1	Address numbers must be plainly legible and visible from the street or road. Where this is not possible, a monument, pole, or other sign must identify the address of the structure at the street or road.
❷		The numbers shall be in contrast to the background and in Arabic writing.
❸		The numbers must be 4″ high with a minimum stroke of ½″.

You Should Know

- Address signage is important because an emergency vehicle must be able to find your building during an emergency.

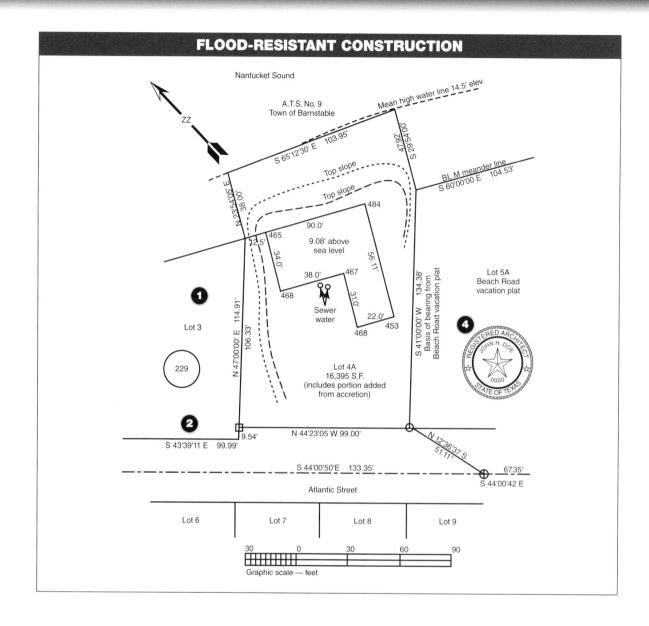

FLOOD-RESISTANT CONSTRUCTION

Nantucket Sound

A.T.S. No. 9
Town of Barnstable

Mean high water line 14.5' elev.

ZZ

S 65'12'30' E   103.95

47.92

S 29'54'00

BLM meander line

S 60'00'00 E   104.53'

N 23'54'05' E   38.00'

Top slope

Top slope

484

90.0'

465   9.08' above
sea level

12.5'

34.0'

56.11'

467

468   38.0'

Sewer
water

31.0'

22.0'

468   453

Lot 5A
Beach Road
vacation plat

S 41'00'00' W   134.38'
Basis of bearing from
Beach Road vacation plat

REGISTERED ARCHITECT
JOHN H. DOE
0000
STATE OF TEXAS

N 47'00'00' E   114.91'

106.33'

Lot 3

229

Lot 4A
16,395 S.F.
(includes portion added
from accretion)

9.54'

N 44'23'05 W 99.00'

N 12'36'37 S
51.11°

S 43'39'11 E   99.99'

S 44'00'50'E   133.35'

67.35'
S 44'00'42 E

Atlantic Street

Lot 6   Lot 7   Lot 8   Lot 9

30   0   30   60   90

Graphic scale — feet

FLOOD-RESISTANT CONSTRUCTION (cont.)

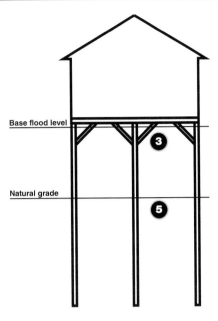

No.	Code	Description
1	R322.1	Buildings in floodways must be designed to meet ASCE 24.
2	R322.1.4	The design flood elevation established through FEMA must be used to define areas prone to flooding.
3	R322.1.8	Building materials and installation methods used for flooring, interior and exterior walls, or wall coverings below the elevation required shall be floor damage-resistant materials that conform to the provisions of FEMA TB2.
4	R322.1.10	As-built documentation is required to be completed by a registered design professional (architect or engineer) for the finish floor elevation height.
5	R322.2.3	Foundation design is affected by flood hazard elevation. If the foundation is within a flood hazard area, it must be designed according to Chapter 4 of the IRC code book.
	R322.3.4	Concrete slabs used for parking, walkways, and similar uses that are subject to undermining or displacement during base flood conditions must be designed to be structurally independent of the foundation system or self-supporting structural slabs capable of remaining intact.

You Should Know

- More stringent requirements for construction occur if a building is within coastal high-hazard areas, including V zones.

FOOTINGS

TABLE R401.4.1 PRESUMPTIVE LOAD-BEARING VALUES OF FOUNDATION MATERIALS[a]

CLASS OF MATERIAL	LOAD-BEARING PRESSURE (pounds per square foot)
Crystalline bedrock	12,000
Sedimentary and foliated rock	4,000
Sandy gravel and/or gravel (GW and GP)	3,000
Sand, silty sand, clayey sand, silty gravel and clayey gravel (SW, SP, SM, SC, GM and GC)	2,000
Clay, sandy clay, silty clay, clayey silt, silt and sandy silt (CL, ML, MH and CH)	1,500[b]

For SI: 1 pound per square foot = 0.0479 kPa.

a. When soil tests are required by Section R401.4, the allowable bearing capacities of the soil shall be part of the recommendations.
b. Where the building official determines that in-place soils with an allowable bearing capacity of less than 1,500 psf are likely to be present at the site, the allowable bearing capacity shall be determined by a soils investigation.

© 2018 IRC®, International Code Council®

Code	Description
R401.4	Where quantifiable data created by accepted soil science methodologies indicate expansive soils, compressible soils, shifting soils, or other questionable soil characteristics are likely to be present, a soil report may be required by the local building department.
Table R401.4.1	Presumptive load-bearing values of foundation materials are given in this table. This allows a designer to classify the soil according to a general category to establish allowable load-bearing pressure.

You Should Know

- A soil report is the basis for foundation design in order that prescriptive requirements may be used.
- Soil bearing capacity is based on the type of material, such as bedrock, sandy gravel, silty sand, silt, or clay. For example: A footing 12″ sq. (1 ft²) could adequately support 1500 pounds when within silty-clay soil.

FOUNDATION MATERIALS

TABLE R402.2 MINIMUM SPECIFIED COMPRESSIVE STRENGTH OF CONCRETE

TYPE OR LOCATION OF CONCRETE CONSTRUCTION	MINIMUM SPECIFIED COMPRESSIVE STRENGTH[a] (f'_c)		
	Weathering Potential[b]		
	Negligible	Moderate	Severe
Basement walls, foundations and other concrete not exposed to the weather	2,500	2,500	2,500[c]
Basement slabs and interior slabs on grade, except garage floor slabs	2,500	2,500	2,500[c]
Basement walls, foundation walls, exterior walls and other vertical concrete work exposed to the weather	2,500	3,000[d]	3,000[d]
Porches, carport slabs and steps exposed to the weather, and garage floor slabs	2,500	3,000[d, e, f]	3,500[d, e, f]

© 2018 IRC®, International Code Council®

For SI: 1 pound per square inch = 6.895 kPa.

a. Strength at 28 days psi.
b. See Table R301.2(1) for weathering potential.
c. Concrete in these locations that may be subject to freezing and thawing during construction shall be air-entrained concrete in accordance with Footnote d.
d. Concrete shall be air-entrained. Total air content (percent by volume of concrete) shall be not less than 5 percent or more than 7 percent.
e. See Section R402.2 for maximum cementitious materials content.
f. For garage floors with a steel-troweled finish, reduction of the total air content (percent by volume of concrete) to not less than 3 percent is permitted if the specified compressive strength of the concrete is increased to not less than 4,000 psi.

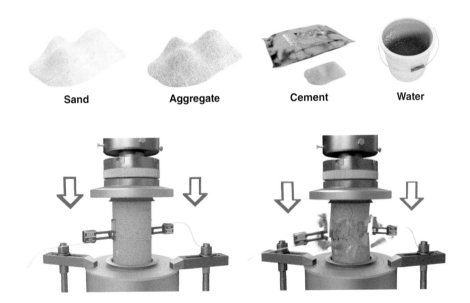

Sand   Aggregate   Cement   Water

FOUNDATION MATERIALS *(cont.)*

Code	Description
R402.2	Concrete used for foundations must have a minimum compressive strength as established in Table R402.2 above.
R402.3	Pre-cast concrete must have a minimum compressive strength of 5000 psi at 28-day cure. Structural reinforcing must have a minimum yield strength of 40,000 psi. Steel reinforcement for pre-cast foundation walls must have a minimum cover of ¾″. A test cylinder evaluates the strength of concrete.
R403.1	Footings must be supported on undisturbed natural soil or engineered fill.

You Should Know

- Other materials such as wood and pre-cast foundation materials may be used according to design strength and imposed loads.

- Compressive strength of concrete is measured by its ability to resist a crushing force in units of pounds per square foot (psi). For example, if concrete has a strength of 3000 psi, it is said to be capable of withstanding a compressive force of 3000 pounds for every 1 in^2 of surface area.

- Concrete is formed by mixing cement, sand, water, and aggregate. Sometimes the mixture is changed to cause the concrete to be stronger or more resistant to environmental effects. Admixtures are anything other than the four basic ingredients. They are sometimes added for various purposes. Accelerators, bonding, plasticizer, retarders, and water reducers are some admixtures.

FOOTINGS

TABLE R403.1(1) MINIMUM WIDTH AND THICKNESS FOR CONCRETE FOOTINGS FOR LIGHT-FRAME CONSTRUCTION (inches)[a, b]

SNOW LOAD OR ROOF LIVE LOAD	STORY AND TYPE OF STRUCTURE WITH LIGHT FRAME	LOAD-BEARING VALUE OF SOIL (psf)					
		1500	2000	2500	3000	3500	4000
20 psf	1 story—slab-on-grade	12 × 6	12 × 6	12 × 6	12 × 6	12 × 6	12 × 6
	1 story—with crawl space	12 × 6	12 × 6	12 × 6	12 × 6	12 × 6	12 × 6
	1 story—plus basement	18 × 6	14 × 6	12 × 6	12 × 6	12 × 6	12 × 6
	2 story—slab-on-grade	12 × 6	12 × 6	12 × 6	12 × 6	12 × 6	12 × 6
	2 story—with crawl space	16 × 6	12 × 6	12 × 6	12 × 6	12 × 6	12 × 6
	2 story—plus basement	22 × 6	16 × 6	13 × 6	12 × 6	12 × 6	12 × 6
	3 story—slab-on-grade	14 × 6	12 × 6	12 × 6	12 × 6	12 × 6	12 × 6
	3 story—with crawl space	19 × 6	14 × 6	12 × 6	12 × 6	12 × 6	12 × 6
	3 story—plus basement	25 × 8	19 × 6	15 × 6	13 × 6	12 × 6	12 × 6
30 psf	1 story—slab-on-grade	12 × 6	12 × 6	12 × 6	12 × 6	12 × 6	12 × 6
	1 story—with crawl space	13 × 6	12 × 6	12 × 6	12 × 6	12 × 6	12 × 6
	1 story—plus basement	19 × 6	14 × 6	12 × 6	12 × 6	12 × 6	12 × 6
	2 story—slab-on-grade	12 × 6	12 × 6	12 × 6	12 × 6	12 × 6	12 × 6
	2 story—with crawl space	17 × 6	13 × 6	12 × 6	12 × 6	12 × 6	12 × 6
	2 story—plus basement	23 × 6	17 × 6	14 × 6	12 × 6	12 × 6	12 × 6
	3 story—slab-on-grade	15 × 6	12 × 6	12 × 6	12 × 6	12 × 6	12 × 6
	3 story—with crawl space	20 × 6	15 × 6	12 × 6	12 × 6	12 × 6	12 × 6
	3 story—plus basement	26 × 8	20 × 6	16 × 6	13 × 6	12 × 6	12 × 6
50 psf	1 story—slab-on-grade	12 × 6	12 × 6	12 × 6	12 × 6	12 × 6	12 × 6
	1 story—with crawl space	16 × 6	12 × 6	12 × 6	12 × 6	12 × 6	12 × 6
	1 story—plus basement	21 × 6	16 × 6	13 × 6	12 × 6	12 × 6	12 × 6
	2 story—slab-on-grade	14 × 6	12 × 6	12 × 6	12 × 6	12 × 6	12 × 6
	2 story—with crawl space	19 × 6	14 × 6	12 × 6	12 × 6	12 × 6	12 × 6
	2 story—plus basement	25 × 7	19 × 6	15 × 6	12 × 6	12 × 6	12 × 6
	3 story—slab-on-grade	17 × 6	13 × 6	12 × 6	12 × 6	12 × 6	12 × 6
	3 story—with crawl space	22 × 6	17 × 6	13 × 6	12 × 6	12 × 6	12 × 6
	3 story—plus basement	28 × 9	21 × 6	17 × 6	14 × 6	12 × 6	12 × 6
70 psf	1 story—slab-on-grade	12 × 6	12 × 6	12 × 6	12 × 6	12 × 6	12 × 6
	1 story—with crawl space	18 × 6	13 × 6	12 × 6	12 × 6	12 × 6	12 × 6
	1 story—plus basement	24 × 7	18 × 6	14 × 6	12 × 6	12 × 6	12 × 6
	2 story—slab-on-grade	16 × 6	12 × 6	12 × 6	12 × 6	12 × 6	12 × 6
	2 story—with crawl space	21 × 6	16 × 6	13 × 6	12 × 6	12 × 6	12 × 6
	2 story—plus basement	27 × 9	20 × 6	16 × 6	14 × 6	12 × 6	12 × 6
	3 story—slab-on-grade	19 × 6	14 × 6	12 × 6	12 × 6	12 × 6	12 × 6
	3 story—with crawl space	25 × 7	18 × 6	15 × 6	12 × 6	12 × 6	12 × 6
	3 story—plus basement	30 × 10	23 × 6	18 × 6	15 × 6	13 × 6	12 × 6

For SI: 1 inch = 25.4 mm, 1 plf = 14.6 N/m. 1 pound per square foot = 47.9 N/m².

a. Interpolation allowed. Extrapolation is not allowed.
b. Based on 32-foot-wide house with load-bearing center wall that carries half of the tributary attic, and floor framing. For every 2 feet of adjustment to the width of the house, add or subtract 2 inches of footing width and 1 inch of footing thickness (but not less than 6 inches thick).

Note: Tables R403.1(1), (2), and (3) have been added to clarify footing requirements for light frame construction, brick veneer, and grouted masonry wall types.

© 2018 IRC ®, International Code Council®

FOOTINGS *(cont.)*

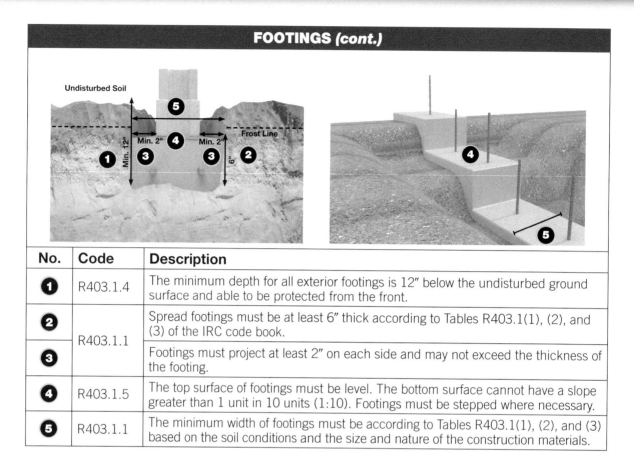

No.	Code	Description
❶	R403.1.4	The minimum depth for all exterior footings is 12″ below the undisturbed ground surface and able to be protected from the front.
❷	R403.1.1	Spread footings must be at least 6″ thick according to Tables R403.1(1), (2), and (3) of the IRC code book.
❸		Footings must project at least 2″ on each side and may not exceed the thickness of the footing.
❹	R403.1.5	The top surface of footings must be level. The bottom surface cannot have a slope greater than 1 unit in 10 units (1:10). Footings must be stepped where necessary.
❺	R403.1.1	The minimum width of footings must be according to Tables R403.1(1), (2), and (3) based on the soil conditions and the size and nature of the construction materials.

You Should Know

- A geotechnical evaluation may be required to determine the soil type and load-bearing value of the soil in order to use the prescriptive Table R403.1.

CONSTRUCTION

ANCHOR BOLTS

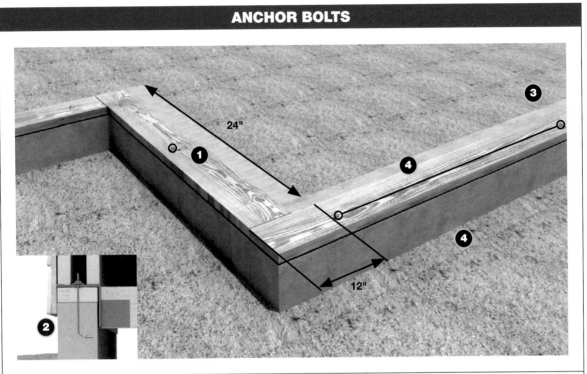

No.	Code	Description
❶		Wood sill plates and walls supported directly on a continuous foundation must be anchored to the foundation.
❷	R403.1.6	Anchor bolts must be at least ½" in diameter and shall extend 7" into concrete or grouted cells of masonry units. A nut and washer shall be tightened on each anchor bolt.
❸		There must be a minimum of two bolts per plate with one / bolt located within 12" or less than 7 bar diameters from the end of the plate section.
❹		Anchor bolts must be spaced not more than 6' on center.
	R403.1.6	Cold-formed steel framing shall be anchored directly to the foundation or fastened to wood sill plates.

You Should Know

- There are alternatives to standard anchor bolts that are manufactured products. If these have been listed and tested and approved in the local jurisdiction, they may be used according to the manufacturer's specifications and installation instructions.

FOOTINGS ON OR ADJACENT TO SLOPES AND FROST-PROTECTED SHALLOW FOUNDATIONS

Figure R403.1.7.1
Foundation Clearances from Slopes
For SI: 1 foot = 304.8 mm

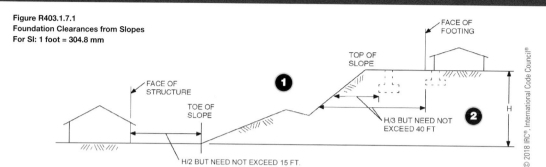

© 2018 IRC®, International Code Council®

③ TABLE R403.3(1) MINIMUM FOOTING DEPTH AND INSULATION REQUIREMENTS FOR FROST-PROTECTED FOOTINGS IN HEATED BUILDINGS[a]

AIR FREEZING INDEX (°F-days)[b]	MINIMUM FOOTING DEPTH, D (inches)	VERTICAL INSULATION R-VALUE[c, d]	HORIZONTAL INSULATION R-VALUE[c, e]		HORIZONTAL INSULATION DIMENSIONS PER FIGURE R403.3(1) (inches)		
			Along walls	At corners	A	B	C
1,500 or less	12	4.5	Not required	Not required	Not required	Not required	Not required
2,000	14	5.6	Not required	Not required	Not required	Not required	Not required
2,500	16	6.7	1.7	4.9	12	24	40
3,000	16	7.8	6.5	8.6	12	24	40
3,500	16	9.0	8.0	11.2	24	30	60
4,000	16	10.1	10.5	13.1	24	36	60

For SI: 1 inch = 25.4 mm, °C = [(°F) − 32]/1.8.

a. Insulation requirements are for protection against frost damage in heated buildings. Greater values may be required to meet energy conservation standards.
b. See Figure R403.3(2) or Table R403.3(2) for Air Freezing Index values.
c. Insulation materials shall provide the stated minimum R-values under long-term exposure to moist, below-ground conditions in freezing climates. The following R-values shall be used to determine insulation thicknesses required for this application: Type II expanded polystyrene-3.2R per inch for vertical insulation and 2.6R per inch for horizontal insulation; Type IX expanded polystyrene-3.4R per inch of vertical insulation and 2.8R for horizontal insulation; Types IV, V, VI, VII, and X extruded polystyrene-4.5R per inch for vertical insulation and 4.0R per inch for horizontal insulation.
d. Vertical insulation shall be expanded polystyrene insulation or extruded polystyrene insulation.
e. Horizontal insulation shall be extruded polystyrene insulation.

© 2018 IRC®, International Code Council®

No.	Code	Description
❶	R403.1.7.1	Buildings below slopes must be at a sufficient distance to provide protection from slope drainage, erosion, and shallow failures.
❷	Figure R403.1.7.1	The face of the structure must be away from the toe of an ascending slope by H/2. The face of a structure must be away from a descending slope by H/3.
	R403.1.7.3	On graded sites, the top of an exterior foundation must be above the elevation of the street gutter or the inlet of an approved drainage device a minimum of 12″ plus 2%.
❸	R403.3	Frost-protected shallow foundations are permitted in some conditions.

You Should Know

- Frost-protected shallow foundations (FPSF) are allowed where the monthly mean temperature is maintained at 65°F.
- These FPSF are not required to extend below the frost zone. Footing depth and insulation required is dependent upon climate conditions. See Table R403.3(1).

FOUNDATION WALLS

TABLE R404.1.1(1) PLAIN MASONRY FOUNDATION WALLS[f]

MAXIMUM WALL HEIGHT (feet)	MAXIMUM UN-BALANCED BACK-FILL HEIGHT[c] (feet)	PLAIN MASONRY[a] MINIMUM NOMINAL WALL THICKNESS (inches) Soil classes[b]		
		GW, GP, SW and SP	GM, GC, SM, SM-SC and ML	SC, MH, ML-CL and inorganic CL
5	4	6 solid[d] or 8	6 solid[d] or 8	6 solid[d] or 8
	5	6 solid[d] or 8	8	10
6	4	6 solid[d] or 8	6 solid[d] or 8	6 solid[d] or 8
	5	6 solid[d] or 8	8	10
	6	8	10	12
7	4	6 solid[d] or 8	8	8
	5	6 solid[d] or 8	10	10
	6	10	12	10 solid[d]
	7	12	10 solid[d]	12 solid[d]
8	4	6 solid[d] or 8	6 solid[d] or 8	8
	5	6 solid[d] or 8	10	12
	6	10	12	12 solid[d]
	7	12	12 solid[d]	Footnote e
	8	10 grout[d]	12 grout[d]	Footnote e
9	4	6 grout[d] or 8 solid[d] or 12	6 grout[d] or 8 solid[d]	8 grout[d] or 10 solid[d]
	5	6 grout[d] or 10 solid[d]	8 grout[d] or 12 solid[d]	8 grout[d]
	6	8 grout[d] or 12 solid[d]	10 grout[d]	10 grout[d]
	7	10 grout[d]	10 grout[d]	12 grout
	8	10 grout[d]	12 grout	Footnote e
	9	12 grout	Footnote e	Footnote e

For SI: 1 inch = 25.4 mm, 1 foot = 304.8 mm, 1 pound per square inch = 6.895 Pa.

a. Mortar shall be Type M or S and masonry shall be laid in running bond. Ungrouted hollow masonry units are permitted except where otherwise indicated.
b. Soil classes are in accordance with the Unified Soil Classification System. Refer to Table R405.1.
c. Unbalanced backfill height is the difference in height between the exterior finish ground level and the lower of the top of the concrete footing that supports the foundation wall or the interior finish ground level. Where an interior concrete slab-on-grade is provided and is in contact with the interior surface of the foundation wall, measurement of the unbalanced backfill height from the exterior finish ground level to the top of the interior concrete slab is permitted.
d. Solid indicates solid masonry unit; grout indicates grouted hollow units.
e. Wall construction shall be in accordance with either Table R404.1.1(2), Table R404.1.1(3), Table R404.1.1(4), or a design shall be provided.
f. The use of this table shall be prohibited for soil classifications not shown.

Code	Description
Table R404.1.1(1)	Plain masonry foundation walls (no reinforcement) are allowed up to 9' high and in some cases with unbalanced backfill up to 9'.
	Plain masonry walls may be either nongrouted or solid (grout-filled). Depending upon the soil class, the walls may have thicknesses from 6" up to 12".
Table R404.1.1(1) footnotes	Depending upon certain conditions, the height or the soil class would remove the design wall from this table and reference another table with reinforcement.

You Should Know

- Mortar must be Type M or S.
- Masonry must be laid in a running bond.
- Soil classes are in accordance with the Unified Soil Classification System.

FOUNDATION WALLS

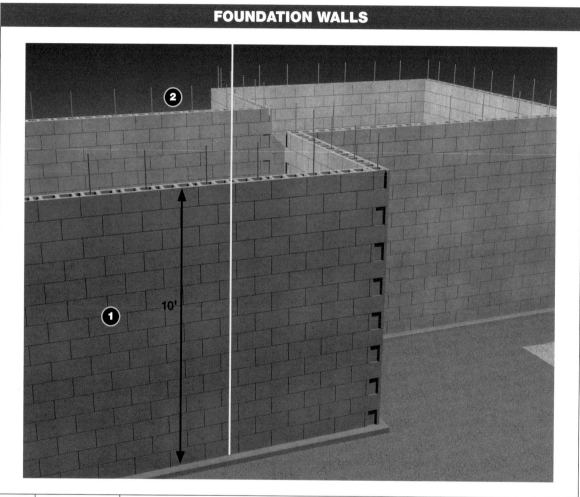

No.	Code	Description
1	Tables R404.1.1(2), (3), and (4)	8"-, 10"-, and 12"-wide masonry foundation walls are permitted with reinforcing up to 10' in height.
2		Required reinforcement is based on the lateral soil load, class of soil present, and the design height.

You Should Know

- Lateral soil load is the horizontal pressure against the foundation wall below grade. Generally, the lateral soil load is distinguished as 30 psf, 45 psf, or 60 psf per foot below grade.
- Reinforcement must be a minimum of grade 60.

CONCRETE FOUNDATION WALLS

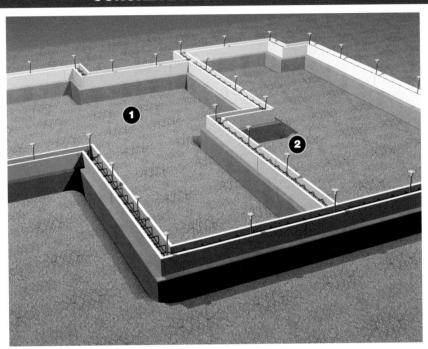

No.	Code	Description
1	R404.1.3.2	Concrete walls up to 10' in height may be designed using the provisions of the IRC.
2	Tables R404.1.2(2), (3), and (4)	Concrete walls may be constructed using the same criteria as insulated concrete forms (ICF). The three tables listed reflect walls that are 6", 8", and 10" thick, respectively. Required reinforcement is based upon soil class, unbalanced backfill height, and lateral soil load.
	Tables R404.1.2(2), (3), and (4)	Vertical reinforcement is required in some conditions and wall heights. The reinforcement ranges from #5 to #6 bars. The size of the bar is based on eighths of an inch. For instance, a #6 bar is ⅝" in diameter and a #5 bar is ⅝" in diameter.
	R404.1.6	Concrete and masonry foundations must extend 4" above the adjacent grade where masonry veneer is used and 6" elsewhere.
	R404.1.9	Isolated masonry piers must be at least 8" in nominal thickness and is limited to a nominal height of not more than 4 times the nominal thickness and a length of not more than 3 times the nominal thickness. Where hollow masonry units are filled with concrete or grout, the piers are permitted to extend to a height of not more than 10 times the nominal thickness.

You Should Know

- Minimum horizontal reinforcement for concrete basement walls of up to 8' in height is a #4 bar within 12" of the top of the wall story and one #4 bar near mid-height of the story. For greater than 8' in height, in addition to within 12" of the top of a wall story, a #4 bar must be installed at third points of the wall story. See Table R404.1.2(1) in the IRC code book.

FOUNDATION DRAINAGE

TABLE R405.1 PROPERTIES OF SOILS CLASSIFIED ACCORDING TO THE UNIFIED SOIL CLASSIFICATION SYSTEM

SOIL GROUP	UNIFIED SOIL CLASSIFICA-TION SYSTEM SYMBOL	SOIL DESCRIPTION	DRAINAGE CHARACTER-ISTICS[a]	FROST HEAVE POTENTIAL	VOLUME CHANGE POTENTIAL EXPANSION[b]
Group I	GW	Well-graded gravels, gravel sand mixtures, little or no fines	Good	Low	Low
	GP	Poorly graded gravels or gravel sand mixtures, little or no fines	Good	Low	Low
	SW	Well-graded sands, gravelly sands, little or no fines	Good	Low	Low
	SP	Poorly graded sands or gravelly sands, little or no fines	Good	Low	Low
	GM	Silty gravels, gravel-sand-silt mixtures	Good	Medium	Low
	SM	Silty sand, sand-silt mixtures	Good	Medium	Low
Group II	GC	Clayey gravels, gravel-sand-clay mixtures	Medium	Medium	Low
	SC	Clayey sands, sand-clay mixture	Medium	Medium	Low
	ML	Inorganic silts and very fine sands, rock flour, silty or clayey fine sands or clayey silts with slight plasticity	Medium	High	Low
	CL	Inorganic clays of low to medium plasticity, gravelly clays, sandy clays, silty clays, lean clays	Medium	Medium	Medium to Low
Group III	CH	Inorganic clays of high plasticity, fat clays	Poor	Medium	High
	MH	Inorganic silts, micaceous or diatomaceous fine sandy or silty soils, elastic silts	Poor	High	High
Group IV	OL	Organic silts and organic silty clays of low plasticity	Poor	Medium	Medium
	OH	Organic clays of medium to high plasticity, organic silts	Unsatisfactory	Medium	High
	Pt	Peat and other highly organic soils	Unsatisfactory	Medium	High

For SI: 1 inch = 25.4 mm.

a. The percolation rate for good drainage is over 4 inches per hour, medium drainage is 2 inches to 4 inches per hour, and poor is less than 2 inches per hour.
b. Soils with a low potential expansion typically have a plasticity index (PI) of 0 to 15, soils with a medium potential expansion have a PI of 10 to 35 and soils with a high potential expansion have a PI greater than 20.

© 2018 IRC®, International Code Council®

No.	Code	Description
	R405	Drains must be provided around all concrete and masonry foundations that retain earth and enclose habitable space located below grade. See exceptions for well-drained soil.
1	R405.2.3	In other than Group I soils, a sump must be provided to drain the porous layer and footing. The sump must be at least 24" in diameter or 20 in² and extend 24" below the bottom of the basement floor.
	R401.3	Surface drainage shall be diverted to a storm sewer conveyance or other approved point of collection that does not create a hazard. Lots shall be graded to drain surface water away from foundation walls. The *grade* shall fall a minimum of 6" within the first 10'. Where lot lines, walls, slopes, or other physical barriers prohibit 6" (152 mm) of fall within 10', drains or swales shall be constructed to ensure drainage away from the structure. Impervious surfaces within 10' of the building foundation shall be sloped a minimum of 2% away from the building.

You Should Know

- Drainage must be to an approved sewer system or to daylight.
- Drainage is not required when the foundation is installed in well-drained soil or sand-gravel mixture soils or Group I soils according to the Unified Soil Classification System.

COLUMNS

The following shall be of naturally durable wood or wood that is preservative-treated:

1. Wood joists or the bottom of a wood structural floor when closer than 18″ or wood girders when closer than 12″ to the exposed ground in crawl spaces or unexcavated area located within the periphery of the building foundation.
2. All wood framing members that rest on concrete or masonry exterior foundation walls and are less than 8″ from the exposed ground.
3. Sills and sleepers on a concrete or masonry slab that is in direct contact with the ground unless separated from such slab by an impervious moisture barrier.
4. The ends of wood girders entering exterior masonry or concrete walls having clearances of less than ½″ on tops, sides, and ends.
5. Wood siding, sheathing, and wall framing on the exterior of a building having a clearance of less than 6″ from the ground or less than 2″ measured vertically from concrete steps, porch slabs, patio slabs, and similar horizontal surfaces exposed to the weather.
6. Wood structural members supporting moisture-permeable floors or roofs that are exposed to the weather, such as concrete or masonry slabs, unless separated from such floors or roofs by an impervious moisture barrier.
7. Wood furring strips or other wood framing members attached directly to the interior of exterior masonry walls or concrete walls below grade except where an approved vapor retarder is applied between the wall and the furring strips or framing members.

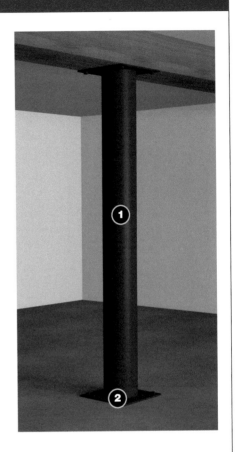

No.	Code	Description
	R407.1	Wood columns must be protected against decay.
❶	R407.2	All surfaces (inside and outside) of steel columns must have a shop coat of rust-inhibitive paint.
❷	R407.3	Columns must be restrained at the bottom end to prevent lateral displacement. Wood columns must be not less than 4″ by 4″. Steel columns must be not less than 3″ in diameter Schedule 40 pipe.

You Should Know

- In seismic zones A, B, and C, columns no more than 48″ in height on a pier or footing are exempt from the requirement for bottom-end lateral displacement.

UNDER-FLOOR SPACE

No.	Code	Description
1	R408.1	Except for space occupied as a basement, under-floor space between the bottom of floor joists and the earth must be ventilated. The ventilation must be a minimum of 1 ft^2 for each 150 ft^2 of under-floor space area. If ground is covered with a Class I vapor retarder material, the net area of ventilation openings may be 1 ft^2 for every 1500 ft^2 of under-floor space.
2	R408.3	Unvented crawl space is permitted where exposed earth is covered with Class I vapor retarder, with joints overlapped by 6″. The edges must (1) extend upward on stem walls at least 6″ or (2) be exposed to either a continuously operated mechanical exhaust ventilation system or a conditioned-air supply delivered to the under-floor area.
3	R408.4	Access is required for under-floor spaces. An opening in the floor must be at least 18″ by 24″. An opening in the exterior wall must be at least 16″ by 24″.

You Should Know

- There are alternatives to providing mechanical ventilation to an unvented crawl space. These are outlined in Section R408.3 in the IRC code book.

UNDER-FLOOR SPACE

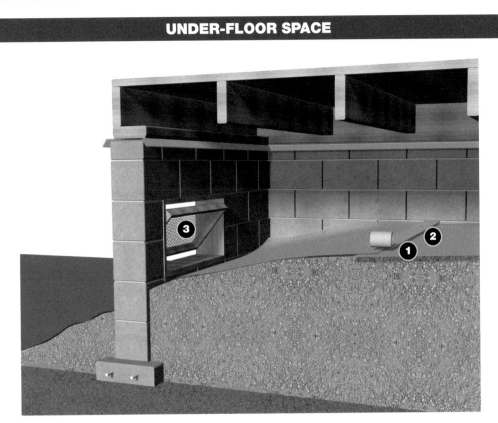

No.	Code	Description
❶	R408.6	The finished grade of an under-floor surface may be located at the bottom of footings unless the groundwater can rise to within 6″ of the finished floor at the building perimeter or the surface water does not drain well from the site. In these cases, the under-floor space must be as high as the outside finished grade unless an approved drainage system is provided.
❷	R408.7	In buildings located in areas prone to flooding, walls enclosing the under-floor space must have flood openings per Section R322.2.2 in the IRC code book. The finished grade level of the under-floor space must be equal to or higher than the outside finished ground level on at least one side.
❸	R322.2.2	Flood vents have specific criteria specified in Section R322.2.2 in the IRC code book.

WOOD FLOOR FRAMING

TABLE R502.3.1(2) FLOOR JOIST SPANS FOR COMMON LUMBER SPECIES (RESIDENTIAL LIVING AREAS, LIVE LOAD = 40 PSF, L/Δ = 360)[b]

JOIST SPACING (inches)	SPECIES AND GRADE		DEAD LOAD = 10 psf				DEAD LOAD = 20 psf			
			2 × 6	2 × 8	2 × 10	2 × 12	2 × 6	2 × 8	2 × 10	2 × 12
			Maximum floor joist spans							
			(ft − in.)	(ft − in.)	(ft − in.)	(ft − in.)	(ft − in.)	(ft − in.)	(ft − in.)	(ft − in.)
12	Douglas fir-larch	SS	11-4	15-0	19-1	23-3	11-4	15-0	19-1	23-3
	Douglas fir-larch	#1	10-11	14-5	18-5	22-0	10-11	14-2	17-4	20-1
	Douglas fir-larch	#2	10-9	14-2	18-0	20-11	10-8	13-6	16-5	19-1
	Douglas fir-larch	#3	8-11	11-3	13-9	16-0	8-1	10-3	12-7	14-7
	Hem-fir	SS	10-9	14-2	18-0	21-11	10-9	14-2	18-0	21-11
	Hem-fir	#1	10-6	13-10	17-8	21-6	10-6	13-10	17-1	19-10
	Hem-fir	#2	10-0	13-2	16-10	20-4	10-0	13-1	16-0	18-6
	Hem-fir	#3	8-8	11-0	13-5	15-7	7-11	10-0	12-3	14-3
	Southern pine	SS	11-2	14-8	18-9	22-10	11-2	14-8	18-9	22-10
	Southern pine	#1	10-9	14-2	18-0	21-11	10-9	14-2	16-11	20-1
	Southern pine	#2	10-3	13-6	16-2	19-1	9-10	12-6	14-9	17-5
	Southern pine	#3	8-2	10-3	12-6	14-9	7-5	9-5	11-5	13-6
	Spruce-pine-fir	SS	10-6	13-10	17-8	21-6	10-6	13-10	17-8	21-6
	Spruce-pine-fir	#1	10-3	13-6	17-3	20-7	10-3	13-3	16-3	18-10
	Spruce-pine-fir	#2	10-3	13-6	17-3	20-7	10-3	13-3	16-3	18-10
	Spruce-pine-fir	#3	8-8	11-0	13-5	15-7	7-11	10-0	12-3	14-3
16	Douglas fir-larch	SS	10-4	13-7	17-4	21-1	10-4	13-7	17-4	21-1
	Douglas fir-larch	#1	9-11	13-1	16-5	19-1	9-8	12-4	15-0	17-5
	Douglas fir-larch	#2	9-9	12-9	15-7	18-1	9-3	11-8	14-3	16-6
	Douglas fir-larch	#3	7-8	9-9	11-11	13-10	7-0	8-11	10-11	12-7
	Hem-fir	SS	9-9	12-10	16-5	19-11	9-9	12-10	16-5	19-11
	Hem-fir	#1	9-6	12-7	16-0	18-10	9-6	12-2	14-10	17-2
	Hem-fir	#2	9-1	12-0	15-2	17-7	8-11	11-4	13-10	16-1
	Hem-fir	#3	7-6	9-6	11-8	13-6	6-10	8-8	10-7	12-4
	Southern pine	SS	10-2	13-4	17-0	20-9	10-2	13-4	17-0	20-9
	Southern pine	#1	9-9	12-10	16-1	19-1	9-9	12-7	14-8	17-5
	Southern pine	#2	9-4	11-10	14-0	16-6	8-6	10-10	12-10	15-1
	Southern pine	#3	7-1	8-11	10-10	12-10	6-5	8-2	9-10	11-8
	Spruce-pine-fir	SS	9-6	12-7	16-0	19-6	9-6	12-7	16-0	19-6
	Spruce-pine-fir	#1	9-4	12-3	15-5	17-10	9-1	11-6	14-1	16-3
	Spruce-pine-fir	#2	9-4	12-3	15-5	17-10	9-1	11-6	14-1	16-3
	Spruce-pine-fir	#3	7-6	9-6	11-8	13-6	6-10	8-8	10-7	12-4
19.2	Douglas fir-larch	SS	9-8	12-10	16-4	19-10	9-8	12-10	16-4	19-6
	Douglas fir-larch	#1	9-4	12-4	15-0	17-5	8-10	11-3	13-8	15-H
	Douglas fir-larch	#2	9-2	11-8	14-3	16-6	8-5	10-8	13-0	15-1
	Douglas fir-larch	#3	7-0	8-11	10-11	12-7	6-5	8-2	9-11	11-6
	Hem-fir	SS	9-2	12-1	15-5	18-9	9-2	12-1	15-5	18-9
	Hem-fir	#1	9-0	11-10	14-10	17-2	8-9	11-1	13-6	15-8
	Hem-fir	#2	8-7	11-3	13-10	16-1	8-2	10-4	12-8	14-8
	Hem-fir	#3	6-10	8-8	10-7	12-4	6-3	7-11	9-8	11-3
	Southern pine	SS	9-6	12-7	16-0	19-6	9-6	12-7	16-0	19-6
	Southern pine	#1	9-2	12-1	14-8	17-5	9-0	11-5	13-5	15-11
	Southern pine	#2	8-6	10-10	12-10	15-1	7-9	9-10	11-8	13-9

(continues)

WOOD FLOOR FRAMING *(cont.)*

TABLE R502.3.1(2) *(cont.)*

JOIST SPACING (inches)	SPECIES AND GRADE		DEAD LOAD = 10 psf				DEAD LOAD = 20 psf			
			2 × 6	2 × 8	2 × 10	2 × 12	2 × 6	2 × 8	2 × 10	2 × 12
			Maximum floor joist spans							
			(ft – in.)	(ft – in.)	(ft – in.)	(ft – in.)	(ft – in.)	(ft – in.)	(ft – in.)	(ft – in.)
19.2	Southern pine	#3	6-5	8-2	9-10	11-8	5-11	7-5	9-0	10-8
	Spruce-pine-fir	SS	9-0	11-10	15-1	18-4	9-0	11-10	15-1	17-9
	Spruce-pine-fir	#	8-9	11-6	14-1	16-3	8-3	10-6	12-10	14-10
	Spruce-pine-fir	#2	8-9	11-6	14-1	16-3	8-3	10-6	12-10	14-10
	Spruce-pine-fir	#3	6-10	8-8	10-7	12-4	6-3	7-11	9-8	11-3
24	Douglas fir-larch	SS	9-0	11-11	15-2	18-5	9-0	11-11	15-0	17-5
	Douglas fir-larch	#1	8-8	11-0	13-5	15-7	7-11	10-0	12-3	14-3
	Douglas fir-larch	#2	8-3	10-5	12-9	14-9	7-6	9-6	11-8	13-6
	Douglas fir-larch	#3	6-3	8-0	9-9	11-3	5-9	7-3	8-11	10-4
	Hem-fir	SS	8-6	11-3	14-4	17-5	8-6	11-3	14-4	16-10a
	Hem-fir	#1	8-4	10-10	13-3	15-5	7-10	9-11	12-1	14-0
	Hem-fir	#2	7-11	10-2	12-5	14-4	7-4	9-3	11-4	13-1
	Hem-fir	#3	6-2	7-9	9-6	11-0	5-7	7-1	8-8	10-1
	Southern pine	SS	8-10	11-8	14-11	18-1	8-10	11-8	14-1 1	18-0
	Southern pine	#1	8-6	11-3	13-1	15-7	8-1	10-3	12-0	14-3
	Southern pine	#2	7-7	9-8	11-5	13-6	7-0	8-10	10-5	12-4
	Southern pine	#3	5-9	7-3	8-10	10-5	5-3	6-8	8-1	9-6
	Spruce-pine-fir	SS	8-4	11-0	14-0	17-0	8-4	11-0	13-8	15-11
	Spruce-pine-fir	#1	8-1	10-3	12-7	14-7	7-5	9-5	11-6	13-4
	Spruce-pine-fir	#2	8-1	10-3	12-7	14-7	7-5	9-5	11-6	13-4
	Spruce-pine-fir	#3	6-2	7-9	9-6	11-0	5-7	7-1	8-8	10-1

For SI: 1 inch = 25.4 mm, 1 foot = 304.8 mm, 1 pound per square foot = 0.0479 kPa.

Note: Check sources for availability of lumber in lengths greater than 20 feet.

a. End bearing length shall be increased to 2 inches.
b. Dead load limits for townhouses in Seismic Design Category C and all structures in Seismic Design Categories D_0, D_1, and D_2 shall be determined in accordance with Section R301.2.2.2.1.

© 2018 IRC®, International Code Council®

Code	Description
R502.3	The allowable span for floor joists must be in accordance with Tables R502.3.1(1) and (2).
R502.3.1	Table R502.3.1(1) is used for sleeping rooms and attics and a live load of 30 psf.
R502.3.2	Table R502.3.1(2) is used for areas other than sleeping rooms and attics. The assumed load is limited to 40 psf.
R502.3.1 and R502.3.2	These tables assume a design dead load not exceeding 20 psf.

You Should Know

- Tables R503.3.2(1) and (2) have been changed to reflect revisions made to span length for Southern Pine, Douglas Fir-Larch, and Hemlock-Fir.
- A similar change appears in revised Table R802.4(1) for ceiling joists spans and rafter spans.

FLOOR FRAMING

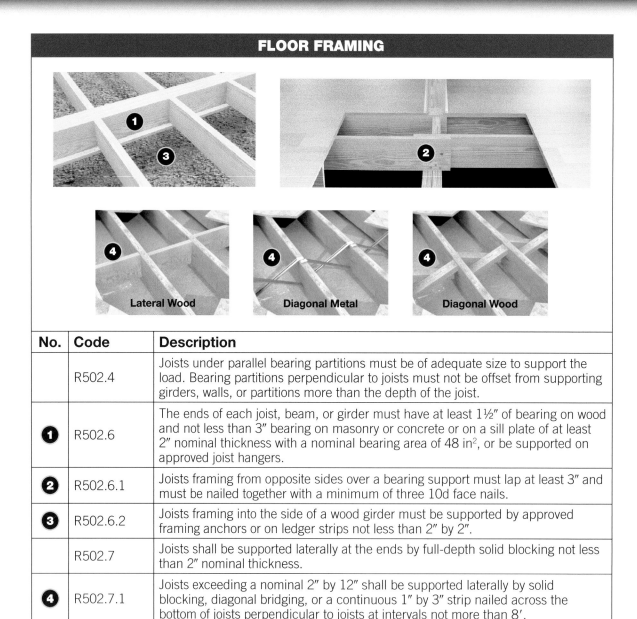

Lateral Wood   Diagonal Metal   Diagonal Wood

No.	Code	Description
	R502.4	Joists under parallel bearing partitions must be of adequate size to support the load. Bearing partitions perpendicular to joists must not be offset from supporting girders, walls, or partitions more than the depth of the joist.
1	R502.6	The ends of each joist, beam, or girder must have at least 1½″ of bearing on wood and not less than 3″ bearing on masonry or concrete or on a sill plate of at least 2″ nominal thickness with a nominal bearing area of 48 in², or be supported on approved joist hangers.
2	R502.6.1	Joists framing from opposite sides over a bearing support must lap at least 3″ and must be nailed together with a minimum of three 10d face nails.
3	R502.6.2	Joists framing into the side of a wood girder must be supported by approved framing anchors or on ledger strips not less than 2″ by 2″.
	R502.7	Joists shall be supported laterally at the ends by full-depth solid blocking not less than 2″ nominal thickness.
4	R502.7.1	Joists exceeding a nominal 2″ by 12″ shall be supported laterally by solid blocking, diagonal bridging, or a continuous 1″ by 3″ strip nailed across the bottom of joists perpendicular to joists at intervals not more than 8′.

You Should Know

- Anywhere the word *approved* is used, it means approved by the jurisdiction. You must ask the local building official if a particular material or method of construction is accepted within the jurisdiction.

FRAMING OF OPENINGS

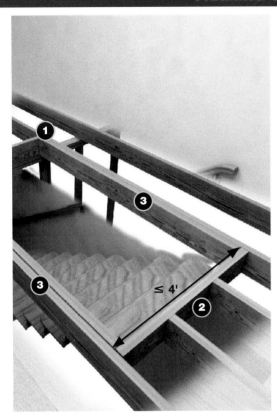 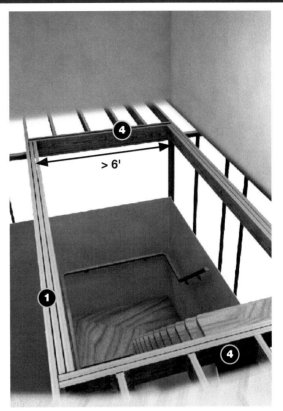

No.	Code	Description
1		Openings in floor framing must be framed with header and trimmer joists.
2		When the header joist does not span more than 4', the header joist may be a single member the same size as the floor joist.
3	R502.10	Single trimmer joists may be used to carry a single header joist that is located within 3' of the trimmer joist bearing.
4		When the header joist span exceeds 4', the trimmer joist and the header joist must be doubled and of sufficient cross section to support the floor joist framing into the header.

CUTTING AND NOTCHING OF FLOOR JOISTS

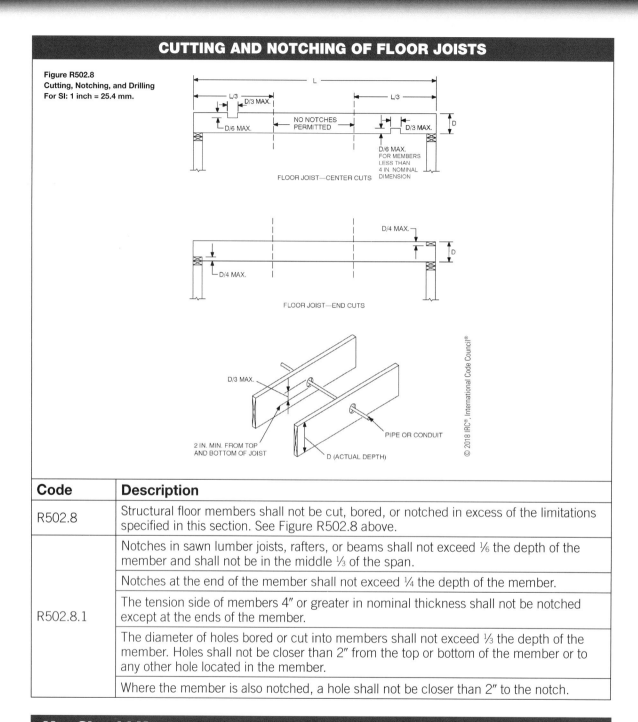

Figure R502.8
Cutting, Notching, and Drilling
For SI: 1 inch = 25.4 mm.

Code	Description
R502.8	Structural floor members shall not be cut, bored, or notched in excess of the limitations specified in this section. See Figure R502.8 above.
R502.8.1	Notches in sawn lumber joists, rafters, or beams shall not exceed ⅙ the depth of the member and shall not be in the middle ⅓ of the span.
	Notches at the end of the member shall not exceed ¼ the depth of the member.
	The tension side of members 4″ or greater in nominal thickness shall not be notched except at the ends of the member.
	The diameter of holes bored or cut into members shall not exceed ⅓ the depth of the member. Holes shall not be closer than 2″ from the top or bottom of the member or to any other hole located in the member.
	Where the member is also notched, a hole shall not be closer than 2″ to the notch.

You Should Know

- Cutting and notching of engineered wood products such as I-Joists and trusses are prohibited except as provided by the manufacturer's recommendations or by a registered design professional.
- Cuts, notches and holes bored in trusses, structural composite lumber, structural glue lam members, cross-laminated timber, or I-Joists is prohibited except where permitted by the manufacturer.

FLOOR SHEATHING

TABLE R503.2.1.1(1) ALLOWABLE SPANS AND LOADS FOR WOOD STRUCTURAL PANELS FOR ROOF AND SUBFLOOR SHEATHING AND COMBINATION SUBFLOOR UNDERLAYMENT[a, b, c]

SPAN RATING	MINIMUM NOMINAL PANEL THICKNESS (inch)	ALLOWABLE LIVE LOAD (psf)[h, l]		MAXIMUM SPAN (inches)		LOAD (pounds per square foot, at maximum span)		MAXIMUM SPAN (inches)
		SPAN @ 16″ o.c.	SPAN @ 24″ o.c.	With edge support[d]	Without edge support	Total load	Live load	
Sheathing[e]				**Roof[f]**				**Subfloor[j]**
16/0	3/8	30	—	16	16	40	30	0
20/0	3/8	50	—	20	20	40	30	0
24/0	3/8	100	30	24	20[g]	40	30	0
24/16	7/16	100	40	24	24	50	40	16
32/16	15/32, 1/2	180	70	32	28	40	30	16[h]
40/20	19/32, 5/8	305	130	40	32	40	30	20[h, i]
48/24	23/32, 3/4	—	175	48	36	45	35	24
60/32	7/8	—	305	60	48	45	35	32
Underlayment, C-C plugged, single floor[e]				**Roof[f]**				**Combination subfloor underlayment[k]**
16 o.c.	19/32, 5/8	100	40	24	24	50	40	16[j]
20 o.c.	19/32, 5/8	150	60	32	32	40	30	20[j, i]
24 o.c.	23/32, 3/4	240	100	48	36	35	25	24
32 o.c.	7/8	—	185	48	40	50	40	32
48 o.c.	1 3/32, 1 1/8	—	290	60	48	50	40	48

For SI: 1 inch = 25.4 mm, 1 pound per square foot = 0.0479 kPa.

a. The allowable total loads were determined using a dead load of 10 psf. If the dead load exceeds 10 psf, then the live load shall be reduced accordingly.
b. Panels continuous over two or more spans with long dimension (strength axis) perpendicular to supports. Spans shall be limited to values shown because of possible effect of concentrated loads.
c. Applies to panels 24 inches or wider.
d. Lumber blocking, panel edge clips (one midway between each support, except two equally spaced between supports when span is 48 inches), tongue-and-groove panel edges, or other approved type of edge support.
e. Includes Structural 1 panels in these grades.
f. Uniform load deflection limitation: 1/180 of span under live load plus dead load, 1/240 of span under live load only.
g. Maximum span 24 inches for 15/32-and 1/2-inch panels.
h. Maximum span 24 inches where 3/4-inch wood finish flooring is installed at right angles to joists.
i. Maximum span 24 inches where 1.5 inches of lightweight concrete or approved cellular concrete is placed over the subfloor.
j. Unsupported edges shall have tongue-and-groove joints or shall be supported with blocking unless minimum nominal 1/4-inch thick underlayment with end and edge joints offset at least 2 inches or 1.5 inches of lightweight concrete or approved cellular concrete is placed over the subfloor, or 3/4-inch wood finish flooring is installed at right angles to the supports. Allowable uniform live load at maximum span, based on deflection of 1/360 of span, is 100 psf.
k. Unsupported edges shall have tongue-and-groove joints or shall be supported by blocking unless nominal 1/4-inch-thick underlayment with end and edge joints offset at least 2 inches or 3/4-inch wood finish flooring is installed at right angles to the supports. Allowable uniform live load at maximum span, based on deflection of 1/360 of span, is 100 psf, except panels with a span rating of 48 on center are limited to 65 psf total uniform load at maximum span.
l. Allowable live load values at spans of 16″ o.c. and 24″ o.c taken from reference standard APA E30, APA Engineered Wood Construction Guide. Refer to reference standard for allowable spans not listed in the table.

FLOOR SHEATHING *(cont.)*

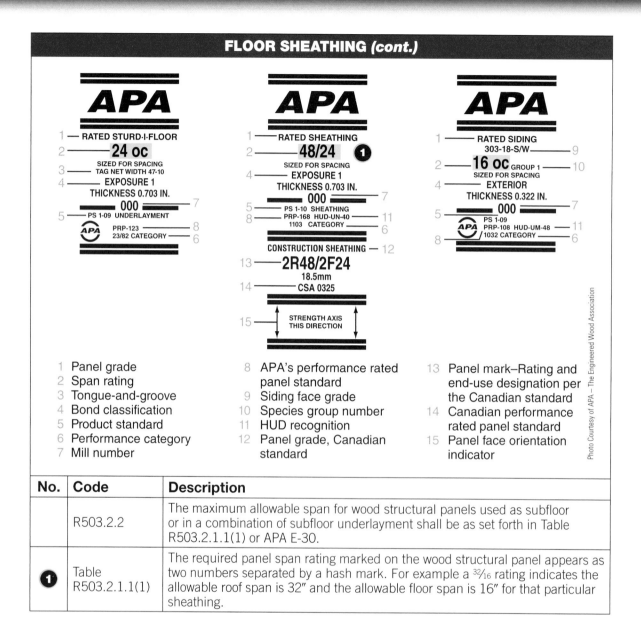

1. Panel grade
2. Span rating
3. Tongue-and-groove
4. Bond classification
5. Product standard
6. Performance category
7. Mill number

8. APA's performance rated panel standard
9. Siding face grade
10. Species group number
11. HUD recognition
12. Panel grade, Canadian standard

13. Panel mark–Rating and end-use designation per the Canadian standard
14. Canadian performance rated panel standard
15. Panel face orientation indicator

Photo Courtesy of APA – The Engineered Wood Association

No.	Code	Description
	R503.2.2	The maximum allowable span for wood structural panels used as subfloor or in a combination of subfloor underlayment shall be as set forth in Table R503.2.1.1(1) or APA E-30.
❶	Table R503.2.1.1(1)	The required panel span rating marked on the wood structural panel appears as two numbers separated by a hash mark. For example a $\frac{32}{16}$ rating indicates the allowable roof span is 32″ and the allowable floor span is 16″ for that particular sheathing.

You Should Know

- Particleboard must conform to ANSI A208.1 and shall be identified by a grade mark or certificate of inspection issued by an *approved agency*.
- The minimum thickness of lumber used as floor sheathing is based on joist or beam spacing and the orientation of the lumber to the joist (perpendicular or diagonal).
- The table illustrated is for both roof and subfloor sheathing. Be sure to distinguish which you are designing. The maximum spans for wood structural panel floor sheathing are limited by the stresses and deflection imposed by the design live loads.

FASTENERS: NAILS AND STAPLES

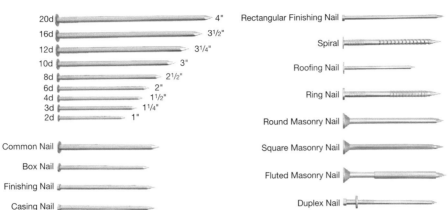

20d	4"	Rectangular Finishing Nail
16d	3½"	Spiral
12d	3¼"	Roofing Nail
10d	3"	Ring Nail
8d	2½"	Round Masonry Nail
6d	2"	Square Masonry Nail
4d	1½"	Fluted Masonry Nail
3d	1¼"	Duplex Nail
2d	1"	
Common Nail		
Box Nail		
Finishing Nail		
Casing Nail		

The nail sizes based on the term *penny* is believed to have first come into use in the early 1600s. The English monetary unit was the pound sterling, which was divided into shillings and pence. The cost of 100 nails in pence in the 1600s is how we refer to nail sizes today. For example, 100 nails that sold for 6 pence were called 6 penny nails (we now refer to them as 6d nails, and 6d is the abbreviation of 6 pence); 100 larger nails that sold for 12 pence were called 12d nails; and so on. Although the price of nails did not standardize the size, the constant established by such led to the standard sizes as we refer to them today.

TYPES OF NAILING

FACE NAILING

The nail is driven directly through the thickness of the member and into the component to which it is being attached.

TOE NAILING

The nail is driven at an angle through the face (thickness) of the lumber and into the component to which it is being attached.

BLIND NAILING

Similar to toe nailing in that the nail is driven at an angle but different in that it will be concealed by the next board (component) to be attached.

FASTENERS: NAILS AND STAPLES *(cont.)*	
Code	**Description**
R502.9	Floor framing shall be nailed in accordance with Table R602.3(1) in the IRC code book.

You Should Know

- Face nailing is driving the nail directly through the face or thickness of the wood member and into the connecting member.
- Toe nailing is driving a nail at an angle through the face of the member and into the connecting member.
- Note that the fastener schedule on the adjoining page serves floor, wall, and roof framing elements.
- Nail sizes are based on the term *penny* and identified with a number and the letter *d*, as in 16d.

FASTENING SCHEDULE

TABLE R602.3(1) FASTENING SCHEDULE

ITEM	DESCRIPTION OF BUILDING ELEMENTS	NUMBER AND TYPE OF FASTENER[a, b, c]	SPACING OF LOCATION
	Roof		
1	Blocking between ceiling joists or rafters to top plate	4-8d box (2½" × 0.113") or 3-8d common (2½" × 0.131"); or 3-10d box (3" × 0.128"); or 3-3" × 0.131" nails	Toe nail
2	Ceiling joists to top plate	4-8d box (2½" × 0.113"); or 3-8d common (2½" × 0.131"); or 3-10d box (3" × 0.128"); or 3-3" × 0.131" nails	Per joist, toe nail
3	Ceiling joist not attached to parallel rafter, laps over partitions [see Sections R802.3.1, R802.3.2 and Table R802.5.l(9)]	4-10d box (3" × 0.128"); or 3-16d common (3½" × 0.162"); or 4-3" × 0.131" nails	Face nail
4	Ceiling joist attached to parallel rafter (heel joint) [see Sections R802.3.1 and R802.3.2 and Table R802.5.l(9)]	Table R802.5.1(9)	Face nail
5	Collar tie to rafter, face nail or 1¼" × 20 ga. ridge strap to rafter	4-10d box (3" × 0.128"); or 3-10d common (3" × 0.148"); or 4-3" × 0.131" nails	Face nail each rafter
6	Rafter or roof truss to plate	3-16d box nails (3½" × 0.135"); or 3-10d common nails (3" × 0.148"); or 4-10d box (3" × 0.128"); or 4-3" × 0.131" nails	2 toe nails on one side and 1 toe nail on opposite side of each rafter or truss'
7	Roof rafters to ridge, valley or hip rafters or roof rafter to minimum 2" ridge beam	4-16d (3½" × 0.135"); or 3-10d common (3½" × 0.148"); or 4-10d box (3" × 0.128"); or 4-3" × 0.131" nails	Toe nail
		3-16d box 3½" × 0.135"); or 2-16d common (3½" × 0.162"); or 3-10d box (3" × 0.128"); or 3-3" × 0.131" nails	End nail
	Wall		
8	Stud to stud (not at braced wall panels)	16d common (3½" × 0.162")	24" o.c. face nail
		10d box (3" × 0.128"); or 3" × 0.131" nails	16" o.c. face nail
9	Stud to stud and abutting studs at intersecting wall corners (at braced wall panels)	16d box (3½" × 0.135"); or 3" × 0.131" nails	12" o.c. face nail
		16d common (3½" × 0.162")	16" o.c. face nail
10	Built-up header (2" to 2" header with ½" spacer)	16d common (3½" × 0.162")	16" o.c. each edge face nail
		16d box (3½" × 0.135")	12" o.c. each edge face nail
11	Continuous header to stud	5-8d box (2½" × 0.113"); or 4-8d common (2½" × 0.131"); or 4-10d box (3" × 0.128")	Toe nail
12	Top plate to top plate	16d common (3½" × 0.162")	16" o.c. face nail
		10d box (3" × 0.128"); or 3" × 0.131" nails	12" o.c. face nail
13	Double top plate splice for SDCs A-D$_2$ with seismic braced wall line spacing < 259	8-16d common (3½" × 0.162"); or 12-16d box (3½" × 0.135"); or 12-10d box (3" × 0.128"); or 12-3" × 0.131" nails	Face nail on each side of end joint (minimum 24" lap splice length each side of end joint)
	Double top plate splice SDCs D$_0$, D$_1$, or D$_2$; and braced wall line spacing ≥ 259	12-16d (3½" × 0.135")	
14	Bottom plate to joist, rim joist, band joist or blocking (not at braced wall panels)	16d common (3½" × 0.162")	16" o.c. face nail
		16d box (3½" × 0.135"); or 3" × 0.131" nails	12" o.c. face nail

(continues)

FASTENING SCHEDULE (cont.)

TABLE R602.3(1) (cont.)

ITEM	DESCRIPTION OF BUILDING MATERIALS	NUMBER AND TYPE OF FASTENER[a, b, c]	SPACING AND LOCATION
15	Bottom plate to joist, rim joist, band joist or blocking (at braced wall panel)	3-16d box (3½" × 0.135"); or 2-16d common (3½" × 0.162"); or 4-3" × 0.131" nails	3 each 16" o.c. face nail 2 each 16" o.c. face nail 4 each 16" o.c. face nail
16	Top or bottom plate to stud	4-8d box (2½" × 0.113"); or 3-16d box (3½" × 0.135"); or 4-8d common (2½" × 0.131"); or 4-10d box (3" × 0.128"); or 4-3" × 0.131" nails	Toe nail
16	Top or bottom plate to stud	3-16d box (3½" × 0.135"); or 2-16d common (3½" × 0.162"); or 3-10d box (3" × 0.128"); or 3-3" × 0.131" nails	End nail
17	Top plates, laps at corners and intersections	3-10d box (3" × 0.128"); or 2-16d common (3½" × 0.162"); or 3-3" × 0.131" nails	Face nail
18	1" brace to each stud and plate	3-8d box (2½" × 0.113"); or 2-8d common (2½" × 0.131"); or 2-10d box (3" × 0.128"); or 2 staples 1¾"	Face nail
19	1" × 6" sheathing to each bearing	3-8d box (2½" × 0.113"); or 2-8d common (2½" × 0.131"); or 2-10d box (3" × 0.128"); or 2 staples, 1" crown, 16 ga., 1¾" long	Face nail
20	1" × 8" and wider sheathing to each bearing	3-8d box (2½" × 0.113"); or 3-8d common (2½" × 0.131"); or 3-10d box (3" × 0.128"); or 3 staples, 1" crown, 16 ga., 1¾" long	Face nail
20	1" × 8" and wider sheathing to each bearing	Wider than 1" × 8" 4-8d box (2½" × 0.113"); or 3-8d common (2½" × 0.131"); or 3-10d box (3" × 0.128"); or 4 staples, 1" crown, 16 ga., 1¾" long	Face nail
Floor			
21	Joist to sill, top plate or girder	4-8d box (2½" × 0.113"); or 3-8d common (2½" × 0.131"); or 3-10d box (3" × 0.128"); or 3-3" × 0.131" nails	Toe nail
22	Rim joist, band joist or blocking to sill or top plate (roof applications also)	8d box (2½" × 0.113")	4" o.c. toe nail
22	Rim joist, band joist or blocking to sill or top plate (roof applications also)	8d common (2½" × 0.131"); or 10d box (3" × 0.128"); or 3" × 0.131" nails	6" o.c. toe nail
23	1" × 6" subfloor or less to each joist	3-8d box (2½" × 0.113"); or 2-8d common (2½" × 0.131"); or 3-10d box (3" × 0.128"); or 2 staples, 1" crown, 16 ga., 1¾" long	Face nail
24	2" subfloor to joist or girder	3-16d box (3½" × 0.135"); or 2-16d common (3½" × 0.162")	Blind and face nail
25	2" planks (plank & beam—floor & roof)	3-16d box (3½" × 0.135"); or 2-16d common (3½" × 0.162")	At each bearing, face nail
26	Band or rim joist to joist	3-16d common (3½" × 0.162") 4-10 box (3" × 0.128"); or 4-3" × 0.131" nails; or 4-3" × 14 ga. staples, ⁷⁄₁₆" crown	End nail

(continues)

FASTENING SCHEDULE (cont.)

TABLE R602.3(1) (cont.)

ITEM	DESCRIPTION OF BUILDING ELEMENTS	NUMBER AND TYPE OF FASTENER[a, b, c]		
27	Built-up girders and beams, 2-inch lumber layers	20d common (4″ × 0.192″); or	Nail each layer as follows: 32″ o.c. at top and bottom and staggered.	
		10d box (3″ × 0.128″); or 3″ × 0.131″ nails	24″ o.c. face nail at top and bottom staggered on opposite sides	
		And: 2-20d common (4″ × 0.192″); or 10d box (3″ × 0.128″); or 3-3″ × 0.131″ nails	Face nail at ends and at each splice	
28	Ledger strip supporting joists or rafters	4-16d box (3½″ × 0.135″); or 3-16d common (3½″ × 0.162″); or 4-10d box (3″ × 0.128″); or 4-3″ × 0.131″ nails	At each joist or rafter, face nail	
29	Bridging or blocking to joist	2-10d (3″ × 0.128″)	Each end, toe nail	

ITEM	DESCRIPTION OF BUILDING ELEMENTS	NUMBER AND TYPE OF FASTENER[a, b, c]	SPACING OF FASTENERS	
			Edges (inches)[h]	Intermediate supports[c,e] (inches)
Wood structural panels, subfloor, roof and interior wall sheathing to framing and particleboard wall sheathing to framing [see Table R602.3(3) for wood structural panel _exterior_ wall sheathing to wall framing]				
30	⅜″ − ½″	6d common (2″ × 0.113″) nail (subfloor, wall)[i] 8d common (2½″ × 0.131″) nail (roof)	6	12[f]
31	¹⁹⁄₃₂″ − 1″	8d common nail (2½″ × 0.131″)	6	12[f]
32	1⅛″ − 1¼″	10d common (3″ × 0.148″) nail; or 8d (2½″ × 0.131″) deformed nail	6	12
Other wall sheathing[g]				
33	½″ structural cellulosic fiber-board sheathing	1½″ galvanized roofing nail, ⁷⁄₁₆″ head diameter, or 1″ crown staple 16 ga., 1¼″ long	3	6
34	²⁵⁄₃₂″ structural cellulosic fiberboard sheathing	1¾″ galvanized roofing nail, ⁷⁄₁₆″ head diameter, or 1″ crown staple 16 ga., 1¼″ long	3	6
35	½″ gypsum sheathing[d]	1½″ galvanized roofing nail; staple galvanized, 1½″ long; 1¼″ screws. Type W or S	7	7
36	⅝″ gypsum sheathing[d]	1¾″ galvanized roofing nail; staple galvanized, 1⅝″ long; 1⅝″ screws. Type W or S	7	7
Wood structural panels, combination subfloor underlayment to framing				
37	¾″ and less	6d deformed (2″ × 0.120″) nail; or 8d common (2½″ × 0.131″) nail	6	12
38	⅞″ − 1″	8d common (2½″ × 0.131″) nail; or 8d deformed (2½″ × 0.120″) nail	6	12
39	1⅛″ − 1¼″	10d common (3″ × 0.148″) nail; or 8d deformed (2½″ × 0.120″) nail	6	12

For SI: 1 inch = 25.4 mm, 1 foot = 304.8 mm, 1 mile per hour = 0.447 m/s; 1 psi = 6.895 MPa.

a. Nails are smooth-common, box or deformed shanks except where otherwise stated. Nails used for framing and sheathing connections shall have minimum average bending yield strengths as shown: 80 ksi for shank diameter of 0.192 inch (20d common nail). 90 ksi for shank diameters larger than 0.142 inch but not larger than 0.177 inch, and 100 ksi for shank diameters of 0.142 inch or less.

b. Staples are 16 gage wire and have a minimum ⁷⁄₁₆-inch on diameter crown width.

c. Nails shall be spaced at not more than 6 inches on center at all supports where spans are 48 inches or greater.

d. Four-foot by 8-foot or 4-foot by 9-foot panels shall be applied vertically.

e. Spacing of fasteners not included in this table shall be based on Table R602.3(2).

f. Where the ultimate design wind speed is 130 mph or less, nails for attaching wood structural panel roof sheathing to gable end wall framing shall be spaced 6 inches on center. Where the ultimate design wind speed is greater than 130 mph. nails for attaching panel roof sheathing to intermediate supports shall be spaced 6 inches on center for minimum 48-inch distance from ridges, eaves and gable end walls; and 4 inches on center to gable end wall framing.

g. Gypsum sheathing shall conform to ASTM C 1396 and shall be installed in accordance with GA 253. Fiberboard sheathing shall conform to ASTM C 208.

h. Spacing of fasteners on floor sheathing panel edges applies to panel edges supported by framing members and required blocking and at floor perimeters only. Spacing of fasteners on roof sheathing panel edges applies to panel edges supported by framing members and required blocking. Blocking of roof or floor sheathing panel edges perpendicular to the framing members need not be provided except as required by other provisions of this code. Floor perimeter shall be supported by framing members or solid blocking.

i. Where a rafter is fastened to an adjacent parallel ceiling joist in accordance with this schedule, provide two toe nails on one side of the rafter and toe nails from the ceiling joist to top plate in accordance with this schedule. The toe nail on the opposite side of the rafter shall not be required.

You Should Know

- While located in the IRC chapter that regulates wall framing, R602.3(1) applies to fastening of wood members in floors, walls, and roof systems.
- Table R602.3(2) in the IRC code book refers to alternative attachments, including staples.
- A single top plate may be used with certain conditions.

CONCRETE FLOORS

No.	Code	Description
1	R506.1	Concrete slab-on-grade floors shall be a minimum of 3½" thick
2	R506.2	The area within the foundation walls shall have all vegetation, topsoil, and foreign matter removed.
3	R506.2.1	Fill materials must be clear of vegetation and foreign matter. The fill must be compacted to assure uniform support of the slab and (except where approved) shall not exceed 24" for clean sand or gravel and 8" for earth.
4	R506.2.2	A 4"-thick base course consisting of clean sand, gravel, crushed stone, crushed concrete, or crushed blast furnace slag passing a 2" sieve shall be placed on the prepared subgrade when the slab is below grade.
5	R506.2.3	With some exceptions, a 6-mil polyethylene or approved vapor retarder with joints lapped 6" must be placed between the concrete floor slab and the base course or on the prepared subgrade where no base course exists.

You Should Know

- Concrete slabs on expansive soils that perform adequately are permitted, subject to approval by the building official.
- Where provided, reinforcement within a slab on grade must remain in place between the center and the upper third of the slab for the duration of concrete placement. Such reinforcement mitigates severe cracking.

EXTERIOR DECKS

② TABLE R507.4 DECK POST HEIGHT[a]

DECK POST SIZE	MAXIMUM HEIGHT[b, b] (feet-inches)
4 × 4	6-9[c]
4 × 6	8
6 × 6	14
8 × 8	14

For SI: 1 inch = 25.4 mm, 1 foot = 304.8 mm, 1 pound per square foot = 0.0479 kPa.
a. Measured to the underside of the beam.
b. Based on 40 psf live load.
c. The maximum permitted height is 8 feet for one-ply and two-ply beams. The maximum permitted height for three-ply beams on post cap is 6 feet 9 inches.

© 2018 IRC®, International Code Council®

EXTERIOR DECKS *(cont.)*

TABLE R507.3.1
MINIMUM FOOTING SIZE FOR DECKS

LIVE / GROUND SNOW LOAD[b] (psf)	TRIBUTARY AREA (sq. ft.)	LOAD BEARING VALUE OF SOILS[a, c, d] (psf)											
		1500[e]			2000[e]			2500[e]			≥ 3000[e]		
		Side of a square footing (inches)	Diameter of a round footing (inches)	Thickness (inches)	Side of a square footing (inches)	Diameter of a round footing (inches)	Thickness (inches)	Side of a square footing (inches)	Diameter of a round footing (inches)	Thickness (inches)	Side of a square footing (inches)	Diameter of a round footing (inches)	Thickness (inches)
40	20	12	14	6	12	14	6	12	14	6	12	14	6
	40	14	16	6	12	14	6	12	14	6	12	14	6
	60	17	19	6	15	17	6	13	15	6	12	14	6
	80	20	22	7	17	19	7	15	17	6	14	16	6
	100	22	25	8	19	21	8	17	19	6	15	17	6
	120	24	27	9	21	23	9	19	21	6	17	19	6
	140	26	29	10	22	25	10	20	23	7	18	21	7
	160	28	31	11	24	27	11	21	24	8	20	22	7
50	20	12	14	6	12	14	6	12	14	6	12	14	6
	40	15	17	6	13	15	6	12	14	6	12	14	6
	60	19	21	6	16	18	6	14	16	6	13	15	6
	80	21	24	8	19	21	8	17	19	6	15	17	6
	100	24	27	9	21	23	9	19	21	6	17	19	6
	120	26	30	10	23	26	10	20	23	7	19	21	6
	140	28	32	11	25	28	11	22	25	8	20	23	7
	160	30	34	12	26	30	12	24	27	9	21	24	8
60	20	12	14	6	12	14	6	12	14	6	12	14	6
	40	16	19	6	14	16	6	13	14	6	12	14	6
	60	20	23	7	17	20	7	16	18	6	14	16	6
	80	23	26	9	20	23	9	18	20	7	16	19	6
	100	26	29	10	22	25	10	20	23	8	18	21	6
	120	28	32	11	25	28	11	22	25	9	20	23	7
	140	31	35	12	27	30	12	24	27	10	22	24	8
	160	33	37	13	28	32	13	25	29	11	23	26	9
70	20	12	14	6	12	14	6	12	14	6	12	14	6
	40	18	20	6	15	17	6	14	15	6	12	14	6
	60	21	24	8	19	21	8	17	19	6	15	17	6
	80	25	28	9	21	24	9	19	22	7	18	20	6
	100	28	31	11	24	27	10	21	24	8	20	22	7
	120	30	34	12	26	30	11	24	27	9	21	24	8
	140	33	37	13	28	32	13	25	29	10	23	26	9
	160	35	40	15	30	34	15	27	31	11	25	28	9

For SI: 1 inch = 25.4 mm, 1 square foot = 0.0929 m², 1 pound per square foot = 0.0479 kPa.

a. Interpolation permitted, extrapolation not permitted.
b. Based on highest load case: Dead + Live or Dead + Snow.
c. Assumes minimum square footing to be 12 inches × 12 inches for 6 × 6 post.
d. If the support is a brick or CMU pier, the footing shall have a minimum 2-inch projection on all sides.
e. Area, in square feet, of deck surface supported by post and footings.

EXTERIOR DECKS (cont.)

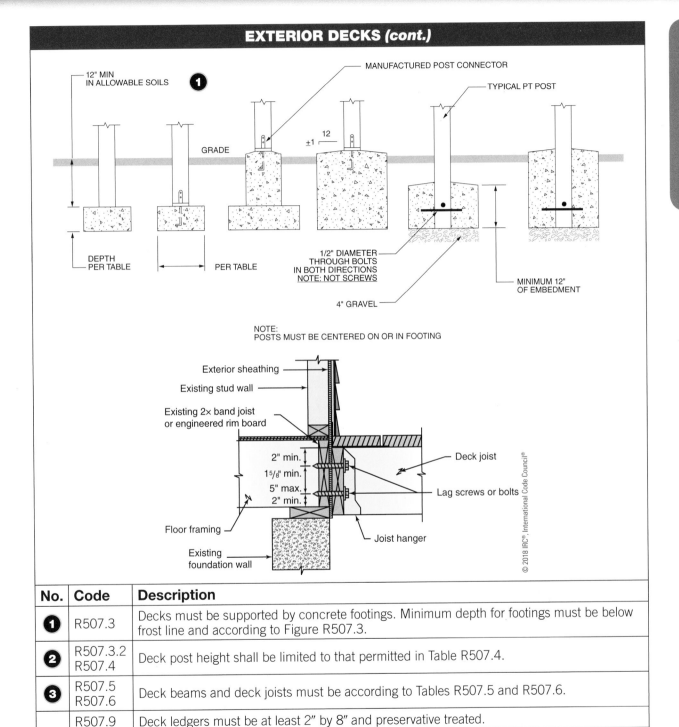

No.	Code	Description
1	R507.3	Decks must be supported by concrete footings. Minimum depth for footings must be below frost line and according to Figure R507.3.
2	R507.3.2 R507.4	Deck post height shall be limited to that permitted in Table R507.4.
3	R507.5 R507.6	Deck beams and deck joists must be according to Tables R507.5 and R507.6.
	R507.9	Deck ledgers must be at least 2″ by 8″ and preservative treated.
	R507.9.1	Band joists supporting ledger must be at least 2″ nominal material (with some exceptions). Band joists must bear fully on primary structure.
	R507.9.2	Vertical and lateral supports for decks must comply with this section. Lateral loads must be transferred to ground or to a structure capable of transmitting them to ground.

EXTERIOR DECKS *(cont.)*

③ TABLE R507.5 DECK BEAM SPAN LENGTHS[a, b, g] (FT. – IN.)

SPECIES[c]	SIZE[d]	DECK JOIST SPAN LESS THAN OR EQUAL TO: (feet)						
		6	8	10	12	14	16	18
Southern pine	$1 - 2 \times 6$	4-11	4-0	3-7	3-3	3-0	2-10	2-8
	$1 - 2 \times 8$	5-11	5-1	4-7	4-2	2-10	3-7	3-5
	$1 - 2 \times 10$	7-0	6-0	5-5	4-11	4-7	4-3	4-0
	$1 - 2 \times 12$	8-3	7-1	6-4	5-10	5-5	5-0	4-9
	$2 - 2 \times 6$	6-11	5-11	5-4	4-10	4-6	4-3	4-0
	$2 - 2 \times 8$	8-9	7-7	6-9	6-2	5-9	5-4	5-0
	$2 - 2 \times 10$	10-4	9-0	8-0	7-4	6-9	6-4	6-0
	$2 - 2 \times 12$	12-2	10-7	9-5	8-7	8-0	7-6	7-0
	$3 - 2 \times 6$	8-2	7-5	6-8	6-1	5-8	5-3	5-0
	$3 - 2 \times 8$	10-10	9-6	8-6	7-9	7-2	6-8	6-4
	$3 - 2 \times 10$	13-0	11-3	10-0	9-2	8-6	7-11	7-6
	$3 - 2 \times 12$	15-3	13-3	11-10	10-9	10-0	9-4	8-10
Douglas fir-larch[e], hem-fir[e], spruce-pine-fir[e], redwood, western cedars, ponderosa pine[f], red pine[f]	3×6 or $2 - 2 \times 6$	5-5	4-8	4-2	3-10	3-6	3-1	2-9
	3×8 or $2 - 2 \times 8$	6-10	5-11	5-4	4-10	4-6	4-1	3-8
	3×10 or $2 - 2 \times 10$	8-4	7-3	6-6	5-11	5-6	5-1	4-8
	3×12 or $2 - 2 \times 12$	9-8	8-5	7-6	6-10	6-4	5-11	5-7
	4×6	6-5	5-6	4-11	4-6	4-2	3-11	3-8
	4×8	8-5	7-3	6-6	5-11	5-6	5-2	4-10
	4×10	9-11	8-7	7-8	7-0	6-6	6-1	5-8
	4×12	11-5	9-11	8-10	8-1	7-6	7-0	6-7
	$3 - 2 \times 6$	7-4	6-8	6-0	5-6	5-1	4-9	4-6
	$3 - 2 \times 8$	9-8	8-6	7-7	6-11	6-5	6-0	5-8
	$3 - 2 \times 10$	12-0	10-5	9-4	8-6	7-10	7-4	6-11
	$3 - 2 \times 12$	13-11	12-1	10-9	9-10	9-1	8-6	8-1

For SI: 1 inch = 25.4 mm, 1 foot = 304.8 mm, 1 pound per square foot = 0.0479 kPa, 1 pound = 0.454 kg.

a. Ground snow load, live load = 40 psf, dead load = 10 psf, L/D = 360 at main span, L/D = 180 at cantilever with a 220-pound point load applied at the end.

b. Beams supporting deck joists from one side only.

c. No. 2 grade, wet service factor.

d. Beam depth shall be greater than or equal to depth of joists with a flush beam condition.

e. Includes incising factor.

f. Northern species. Incising factor not included.

g. Beam cantilevers are limited to the adjacent beam's span divided by 4.

EXTERIOR DECKS *(cont.)*

③ TABLE R507.6 DECK JOIST SPANS FOR COMMON LUMBER SPECIES (FT. – IN.)

SPECIES[a]	SIZE	ALLOWABLE JOIST SPAN[b]			MAXIMUM CANTILEVER[c, f]		
		SPACING OF DECK JOISTS (inches)			SPACING OF DECK JOISTS WITH CANTILEVERS[c] (inches)		
		12	16	24	12	16	24
Southern pine	2 × 6	9-11	9-0	7-7	1-3	1-4	1-6
	2 × 8	13-1	11-10	9-8	2-1	2-3	2-5
	2 × 10	16-2	14-0	11-5	3-4	3-6	2-10
	2 × 12	18-0	16-6	13-6	4-6	4-2	3-4
Douglas fir-larch[d], hem-fir[d] spruce-pine-fir[d],	2 × 6	9-6	8-8	7-2	1-2	1-3	1-5
	2 × 8	12-6	11-1	9-1	1-11	2-1	2-3
	2 × 10	15-8	13-7	11-1	3-1	3-5	2-9
	2 × 12	18-0	15-9	12-10	4-6	3-11	3-3
Redwood, western cedars, ponderosa pine[e], red pine[e]	2 × 6	8-10	8-0	7-0	1-0	1-1	1-2
	2 × 8	11-8	10-7	8-8	1-8	1-10	2-0
	2 × 10	14-11	13-0	10-7	2-8	2-10	2-8
	2 × 12	17-5	15-1	12-4	3-10	3-9	3-1

© 2018 IRC® International Code Council®

For SI: 1 inch = 25.4 mm, 1 foot = 304.8 mm, 1 pound per square foot = 0.0479 kPa, 1 pound = 0.454 kg.

a. No. 2 grade with wet service factor.

b. Ground snow load, live load = 40 psf, dead load = 10 psf, L/Δ = 360.

c. Ground snow load, live load = 40 psf, dead load = 10 psf, L/Δ 360 at main span, L/Δ = 180 at cantilever with a 220-pound point load applied to end.

d. Includes incising factor.

e. Northern species with no incising factor.

f. Cantilevered spans not exceeding the nominal depth of the joist are permitted.

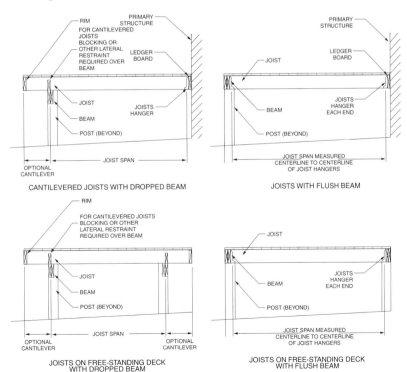

CANTILEVERED JOISTS WITH DROPPED BEAM

JOISTS WITH FLUSH BEAM

JOISTS ON FREE-STANDING DECK WITH DROPPED BEAM

JOISTS ON FREE-STANDING DECK WITH FLUSH BEAM

WALL FRAMING

① TABLE R602.3(5) SIZE, HEIGHT AND SPACING OF WOOD STUDS[a]

STUD SIZE (inches)	BEARING WALLS					NONBEARING WALLS	
	Laterally unsupported stud height[a] (feet)	Maximum spacing when supporting a roof-ceiling assembly or a habitable attic assembly, only (inches)	Maximum spacing when supporting one floor, plus a roof-ceiling assembly or a habitable attic assembly (inches)	Maximum spacing when supporting two floors, plus a roof-ceiling assembly or a habitable attic assembly (inches)	Maximum spacing when supporting one floor height[a] (inches)	Laterally unsupported stud height[a] (feet)	Maximum spacing (inches)
2 × 3[b]	—	—	—	—	—	10	16
2 × 4	10	24[c]	16[c]	—	24	14	24
3 × 4	10	24	24	16	24	14	24
2 × 5	10	24	24	—	24	16	24
2 × 6	10	24	24	16	24	20	24

For SI: 1 inch = 25.4 mm, 1 foot = 304.8 mm.

a. Listed heights are distances between points of lateral support placed perpendicular to the plane of the wall. Bearing walls shall be sheathed on not less than one side or bridging shall be installed not greater than 4 feet apart measured vertically from either end of the stud. Increases in unsupported height are permitted where in compliance with Exception 2 of Section R602.3.1 or designed in accordance with accepted engineering practice.

b. Shall not be used in exterior walls.

c. A habitable attic assembly supported by 2 × 4 studs is limited to a roof span of 32 feet. Where the roof span exceeds 32 feet, the wall studs shall be increased to 2 × 6 or the studs shall be designed in accordance with accepted engineering practice.

TABLE R602.3(6)
ALTERNATE WOOD BEARING WALL STUD SIZE, HEIGHT AND SPACING

STUD HEIGHT	SUPPORTING	STUD SPACING[a]	ULTIMATE DESIGN WIND SPEED					
			115 mph		130 mph[b]		140 mph[b]	
			Maximum roof/floor span[c]		Maximum roof/floor span[c]		Maximum roof/floor span[c]	
			12 ft.	24 ft.	12 ft.	24 ft.	12 ft.	24 ft.
11 ft.	Roof Only	12 in.	2 × 4	2 × 4	2 × 4	2 × 4	2 × 4	2 × 4
		16 in.	2 × 4	2 × 4	2 × 4	2 × 6	2 × 4	2 × 6
		24 in.	2 × 6	2 × 6	2 × 6	2 × 6	2 × 6	2 × 6
	Roof and One Floor	12 in.	2 × 4	2 × 6	2 × 4	2 × 6	2 × 4	2 × 6
		16 in.	2 × 6	2 × 6	2 × 6	2 × 6	2 × 6	2 × 6
		24 in.	2 × 6	2 × 6	2 × 6	2 × 6	2 × 6	2 × 6
12 ft.	Roof Only	12 in	2 × 4	2 × 4	2 × 4	2 × 6	2 × 4	2 × 6
		16 in.	2 × 4	2 × 6	2 × 6	2 × 6	2 × 6	2 × 6
		24 in.	2 × 6	2 × 6	2 × 6	2 × 6	2 × 6	2 × 6
	Roof and One Floor	12 in	2 × 4	2 × 6	2 × 6	2 × 6	2 × 6	2 × 6
		16 in.	2 × 6	2 × 6	2 × 6	2 × 6	2 × 6	2 × 6
		24 in.	2 × 6	2 × 6	2 × 6	2 × 6	2 × 6	DR

For SI:1 inch = 25.4mm, 1 foot = 304.8 mm, 1 mph = 0.447 m/s, 1 pound = 4.448 N.

DR = Design Required.

a. Wall studs not exceeding 16 inches on center shall be sheathed with minimum $1/2$-inch gypsum board on the interior and $3/8$-inch wood structural panel sheathing on the exterior. Wood structural panel sheathing shall be attached with 8d (2.5″ × 0.131″) nails not greater than 6 inches on center along panel edges and 12 inches on center at intermediate supports, and all panel joints shall occur over studs or blocking.

b. Where the ultimate design wind speed exceeds 115 mph, studs shall be attached to top and bottom plates with connectors having a minimum 300-pound lateral capacity.

c. The maximum span is applicable to both single- and multiple-span roof and floor conditions. The roof assembly shall not contain a habitable attic.

WALL FRAMING (cont.)

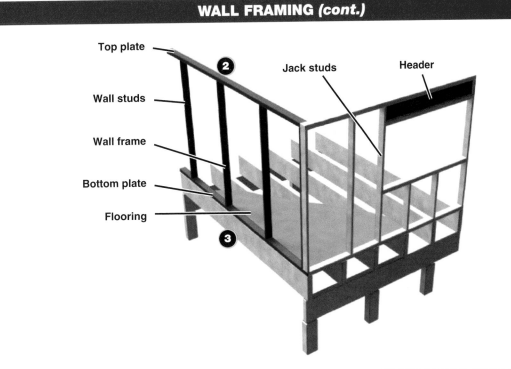

Top plate
Jack studs
Header
Wall studs
Wall frame
Bottom plate
Flooring

No.	Code	Description
❶	R602.3.1	The size, height, and spacing of studs must be according to Table R602.3(5) or as an alternate Table R602.3(6).
❷	R602.3.2	Wood stud walls must be capped with a double top plate installed to provide overlapping at corners and intersections with bearing partitions. End joints in top plates must be overlapped at least 24″. Joints need not occur over studs. Plates shall be not less than 2″ nominal thickness and have a width equal to the studs.
❸	R602.3.4	Studs shall have a full bearing on a nominal 2 by or larger plate or sill having a width equal to the width of the studs. A single top plate may be used with certain conditions.

You Should Know

- Utility-grade studs shall not be spaced more than 16″ on center and shall not support more than a roof and ceiling (not a floor load).
- Studs more than 10′ in height are permitted if they comply with Table R602.3(6) for size of framing members and center spacing of studs if the building is exposed to wind speeds of 100 mph or less.
- Laps at corners and intersections must be face-nailed with two 10d nails.

DRILLING AND NOTCHING OF STUDS

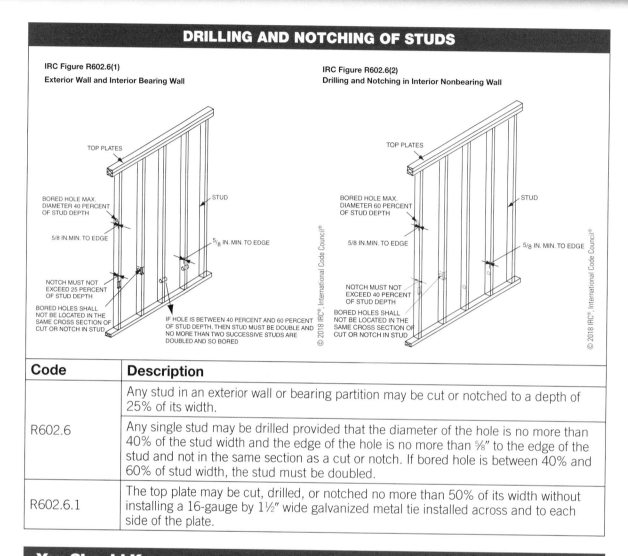

IRC Figure R602.6(1)

Exterior Wall and Interior Bearing Wall

TOP PLATES

BORED HOLE MAX. DIAMETER 40 PERCENT OF STUD DEPTH

STUD

5/8 IN.MIN. TO EDGE

5/8 IN. MIN. TO EDGE

NOTCH MUST NOT EXCEED 25 PERCENT OF STUD DEPTH

BORED HOLES SHALL NOT BE LOCATED IN THE SAME CROSS SECTION OF CUT OR NOTCH IN STUD

IF HOLE IS BETWEEN 40 PERCENT AND 60 PERCENT OF STUD DEPTH, THEN STUD MUST BE DOUBLE AND NO MORE THAN TWO SUCCESSIVE STUDS ARE DOUBLED AND SO BORED

© 2018 IRC®, International Code Council®

IRC Figure R602.6(2)

Drilling and Notching in Interior Nonbearing Wall

TOP PLATES

BORED HOLE MAX. DIAMETER 60 PERCENT OF STUD DEPTH

STUD

5/8 IN.MIN. TO EDGE

5/8 IN. MIN. TO EDGE

NOTCH MUST NOT EXCEED 40 PERCENT OF STUD DEPTH

BORED HOLES SHALL NOT BE LOCATED IN THE SAME CROSS SECTION OF CUT OR NOTCH IN STUD

© 2018 IRC®, International Code Council®

Code	Description
R602.6	Any stud in an exterior wall or bearing partition may be cut or notched to a depth of 25% of its width.
	Any single stud may be drilled provided that the diameter of the hole is no more than 40% of the stud width and the edge of the hole is no more than ⅝" to the edge of the stud and not in the same section as a cut or notch. If bored hole is between 40% and 60% of stud width, the stud must be doubled.
R602.6.1	The top plate may be cut, drilled, or notched no more than 50% of its width without installing a 16-gauge by 1½" wide galvanized metal tie installed across and to each side of the plate.

You Should Know

- Notching and boring hole limitations for wall framing are similar to those for floor and roof framing.
- Notching of studs in interior nonbearing walls may not exceed 40% of stud depth.
- Bored holes in studs in interior nonbearing walls may not exceed 60% of stud depth and may not be within ⅝" of the stud edge.

HEADERS AND GIRDERS

TABLE R602.7(1) GIRDER SPANS[a] AND HEADER SPANS[a] FOR EXTERIOR BEARING WALLS (MAXIMUM SPANS FOR DOUGLAS FIR-LARCH, HEM-FIR, SOUTHERN PINE AND SPRUCE-PINE-FIR[b] AND REQUIRED NUMBER OF JACK STUDS)

GIRDERS AND HEADERS SUPPORTING	SIZE	GROUND SNOW LOAD (psf)[e]																	
		30						50						70					
		Building width[c] (feet)																	
		12		24		36		12		24		36		12		24		36	
		Span[f]	NJ[d]	Span[f]	NJ[d]	Span[f]	NJ[d]	Span[f]	NJ[d]	Span[f]	NJ[d]	Span[f]	NJ[d]	Span[f]	NJ[d]	Span[f]	NJ[d]	Span[f]	NJ[d]
Roof and ceiling	1-2×6	4-0	1	3-1	2	2-7	2	3-5	1	2-8	2	2-3	2	3-0	2	2-4	2	2-0	2
	1-2×8	5-1	2	3-11	2	3-3	2	4-4	2	3-4	2	2-10	2	3-10	2	3-0	2	2-6	3
	1-2×10	6-0	2	4-8	2	3-11	2	5-2	2	4-0	2	3-4	3	4-7	2	3-6	3	3-0	3
	1-2×12	7-1	2	5-5	2	4-7	3	6-1	2	4-8	3	3-11	3	5-5	2	4-2	3	3-6	3
	2-2×4	4-0	1	3-1	1	2-7	1	3-5	1	2-7	1	2-2	1	3-0	1	2-4	1	2-0	1
	2-2×6	6-0	1	4-7	1	3-10	2	5-1	1	3-11	1	3-3	2	4-6	1	3-6	2	2-11	2
	2-2×8	7-7	1	5-9	1	4-10	2	6-5	1	5-0	2	4-2	2	5-9	1	4-5	2	3-9	2
	2-2×10	9-0	1	6-10	2	5-9	2	7-8	2	5-11	2	4-11	2	6-9	2	5-3	2	4-5	2
	2-2×12	10-7	1	8-1	2	6-10	2	9-0	2	6-11	2	5-10	2	8-0	2	6-2	2	5-2	3
	3-2×8	9-5	1	7-3	1	6-1	1	8-1	1	6-3	1	5-3	2	7-2	1	5-6	2	4-8	2
	3-2×10	11-3	1	8-7	1	7-3	2	9-7	1	7-4	2	6-2	2	8-6	1	6-7	2	5-6	2
	3-2×12	13-2	1	10-1	1	8-6	2	11-3	2	8-8	2	7-4	2	10-0	2	7-9	2	6-6	2
	4-2×8	10-11	1	8-4	1	7-0	1	9-4	1	7-2	1	6-0	1	8-3	1	6-4	1	5-4	2
	4-2×10	12-11	1	9-11	1	8-4	1	11-1	1	8-6	1	7-2	2	9-10	1	7-7	2	6-4	2
	4-2×12	15-3	1	11-8	1	9-10	2	13-0	1	10-0	1	8-5	2	11-7	1	8-11	2	7-6	2
Roof, ceiling and one center-bearing floor	1-2×6	3-3	2	2-7	2	2-2	2	3-0	2	2-4	2	2-0	2	2-9	2	2-2	2	1-10	2
	1-2×8	4-1	2	3-3	2	2-9	2	3-9	2	3-0	2	2-6	3	3-6	2	2-9	2	2-4	2
	1-2×10	4-11	2	3-10	2	3-3	2	4-6	2	3-6	3	3-0	3	4-1	2	3-3	3	2-9	3
	1-2×12	5-9	2	4-6	3	3-10	3	5-3	2	4-2	3	3-6	3	4-10	2	3-10	3	3-3	4
	2-2×4	3-3	1	2-6	1	2-2	1	3-0	1	2-4	1	2-0	1	2-8	1	2-2	1	1-10	1
	2-2×6	4-10	1	3-9	2	3-3	2	4-5	2	3-6	2	2-11	2	4-1	2	3-3	2	2-9	2
	2-2×8	6-1	2	4-10	2	4-1	2	5-7	2	4-5	2	3-9	2	5-2	2	4-1	2	3-6	2
	2-2×10	7-3	2	5-8	2	4-10	2	6-8	2	5-3	2	4-5	2	6-1	2	4-10	2	4-1	2
	2-2×12	8-6	2	6-8	2	5-8	2	7-10	2	6-2	2	5-3	2	7-2	2	5-8	2	4-10	3
	3-2×8	7-8	1	6-0	2	5-1	2	7-0	2	5-6	2	4-8	2	6-5	2	5-1	2	4-4	2
	3-2×10	9-1	1	7-2	2	6-1	2	8-4	2	6-7	2	5-7	2	7-8	2	6-1	2	5-2	2
	3-2×12	10-8	2	8-5	2	7-2	2	9-10	2	7-8	2	6-7	2	9-0	2	7-1	2	6-1	2
	4-2×8	8-10	1	6-11	2	5-11	2	8-1	1	6-4	2	5-5	2	7-5	1	5-11	2	5-0	2
	4-2×10	10-6	1	8-3	2	7-0	2	9-8	2	7-7	2	6-5	2	8-10	2	7-0	2	6-0	2
	4-2×12	12-4	1	9-8	2	8-3	2	11-4	2	8-11	2	7-7	2	10-4	2	8-3	2	7-0	2

HEADER, TYP.

ROOF AND CEILING

ROOF, CEILING AND ONE FLOOR (CENTER BEARING)

(continues)

HEADERS AND GIRDERS (cont.)

TABLE R602.7(1) (cont.)

GIRDERS AND HEADERS SUPPORTING	SIZE	GROUND SNOW LOAD (psf)e																
		30						50						70					
		Building width (feet)c																
		12		24		36		12		24		36		12		24		36	
		Span^f	NJ^d	Span^f	NJ^d	Span^f	NJ^d	Span^f	NJ^d	Span^f	NJ^d	Span^f	NJ^d	Span^f	NJ^d	Span^f	NJ^d	Span^f	NJ^d
Roof, ceiling and one clear-span floor	1-2 × 6	2-11	2	2-3	2	1-11	2	2-9	2	2-1	2	1-9	2	2-7	2	2-0	2	1-8	2
	1-2 × 8	3-9	2	2-10	2	2-5	3	3-6	2	2-8	2	2-3	3	3-3	2	2-6	3	2-2	3
	1-2 × 10	4-5	2	3-5	3	2-10	3	4-2	2	3-2	3	2-8	3	3-11	2	3-0	3	2-6	3
	1-2 × 12	5-2	2	4-0	3	3-4	3	4-10	3	3-9	3	3-2	4	4-7	3	3-6	3	3-0	4
	2-2 × 4	2-11	1	2-3	1	1-10	1	2-9	1	2-1	1	1-9	1	2-7	1	2-0	1	1-8	1
	2-2 × 6	4-4	1	3-4	2	2-10	2	4-1	1	3-2	2	2-8	2	3-10	1	3-0	2	2-6	2
	2-2 × 8	5-6	2	4-3	2	3-7	2	5-2	2	4-0	2	3-4	2	4-10	2	3-9	2	3-2	2
	2-2 × 10	6-7	2	5-0	2	4-2	2	6-1	2	4-9	2	4-0	2	5-9	2	4-5	2	3-9	3
	2-2 × 12	7-9	2	5-11	2	4-11	3	7-2	2	5-7	2	4-8	3	6-9	2	5-3	3	4-5	3
	3-2 × 8	6-11	1	5-3	2	4-5	2	6-5	1	5-0	2	4-2	2	6-1	1	4-8	2	4-0	2
ROOF, CEILING AND ONE FLOOR (CLEAR SPAN)	3-2 × 10	8-3	2	6-3	2	5-3	2	7-8	2	5-11	2	5-0	2	7-3	2	5-7	2	4-8	2
	3-2 × 12	9-8	2	7-5	2	6-2	2	9-0	2	7-0	2	5-10	2	8-6	2	6-7	2	5-6	3
	4-2 × 8	8-0	1	6-1	1	5-1	2	7-5	1	5-9	1	4-10	2	7-0	1	5-5	2	4-7	2
	4-2 × 10	9-6	1	7-3	2	6-1	2	8-10	1	6-10	1	5-9	2	8-4	1	6-5	2	5-5	2
	4-2 × 12	11-2	2	8-6	2	7-2	2	10-5	2	8-0	2	6-9	2	9-10	2	7-7	2	6-5	2

(continues)

HEADERS AND GIRDERS (cont.)

TABLE R602.7(1) (cont.)

GIRDERS AND HEADERS SUPPORTING	SIZE	GROUND SNOW LOAD (psf)[e] 30 — Building width[c] (feet) 20 Span[f]	NJ[d]	24 Span[f]	NJ[d]	36 Span[f]	NJ[d]	50 — 20 Span[f]	NJ[d]	24 Span[f]	NJ[d]	36 Span[f]	NJ[d]	70 — 20 Span[f]	NJ[d]	24 Span[f]	NJ[d]	36 Span[f]	NJ[d]
Roof, ceiling and two center-bearing floors	1-2 × 6	2-8	2	2-1	2	1-10	2	2-7	2	2-0	2	1-9	2	2-5	2	1-11	2	1-8	2
	1-2 × 8	3-5	2	2-8	2	2-4	3	3-3	2	2-7	2	2-2	3	3-1	2	2-5	3	2-1	3
	1-2 × 10	4-0	2	3-2	3	2-9	3	3-10	2	3-1	3	2-7	3	3-8	2	2-11	3	2-5	3
	1-2 × 12	4-9	3	3-9	3	3-2	4	4-6	3	3-7	3	3-1	4	4-3	3	3-5	3	2-11	4
	2-2 × 4	2-8	1	2-1	2	1-9	2	2-6	1	2-0	2	1-8	2	2-5	1	1-11	1	1-7	1
	2-2 × 6	4-0	1	3-2	2	2-8	2	3-9	2	3-0	2	2-7	2	3-7	2	2-10	2	2-5	2
	2-2 × 8	5-0	2	4-0	2	3-5	2	4-10	2	3-10	2	3-3	2	4-7	2	3-7	2	3-1	2
	2-2 × 10	6-0	2	4-9	2	4-0	2	5-8	2	4-6	2	3-10	3	5-5	2	4-3	2	3-8	3
	2-2 × 12	7-0	2	5-7	2	4-9	3	6-8	2	5-4	2	4-6	3	6-4	3	5-0	3	4-3	3
	3-2 × 8	6-4	1	5-0	2	4-3	2	6-0	2	4-9	2	4-1	2	5-8	2	4-6	2	3-10	2
ROOF, CEILING AND TWO FLOORS (CENTER BEARING)	3-2 × 10	7-6	2	5-11	2	5-1	2	7-1	2	5-8	2	4-10	2	6-9	2	5-4	2	4-7	2
	3-2 × 12	8-10	2	7-0	2	5-11	2	8-5	2	6-8	2	5-8	3	8-0	2	6-4	2	5-4	3
	4-2 × 8	7-3	1	5-9	1	4-11	2	6-11	1	5-6	2	4-8	2	6-7	1	5-2	2	4-5	2
	4-2 × 10	8-8	1	6-10	2	5-10	2	8-3	2	6-6	2	5-7	2	7-10	2	6-2	2	5-3	2
	4-2 × 12	10-2	2	8-1	2	6-10	2	9-8	2	7-8	2	6-7	2	9-2	2	7-3	2	6-2	2
Roof, ceiling, and two clear-span floors	1-2 × 6	2-3	2	1-9	2	1-5	3	2-3	2	1-8	2	1-4	3	2-2	2	1-8	2	1-4	3
	1-2 × 8	3-4	2	2-6	2	2-2	3	3-4	2	2-6	2	2-2	3	3-2	2	2-6	3	2-1	3
	1-2 × 10	4-3	2	3-3	3	2-8	3	4-3	2	3-3	3	2-8	3	3-9	3	3-0	3	2-8	3
	1-2 × 12	5-0	3	3-10	3	3-2	4	4-10	3	3-10	3	3-2	4	4-10	3	3-9	3	3-2	4
	2-2 × 6	3-4	2	2-6	2	2-2	2	3-4	2	2-6	2	2-2	2	3-3	2	2-6	2	2-1	3
	2-2 × 8	4-3	3	3-3	3	2-8	3	4-3	3	3-3	3	2-8	3	3-11	3	3-0	3	2-8	3
	2-2 × 10	5-0	3	4-0	3	3-2	3	5-0	3	4-0	3	3-2	3	4-8	4	3-9	4	3-2	3
	2-2 × 12	5-11	3	4-9	3	3-11	3	5-8	3	4-8	3	3-11	3	5-6	4	4-8	3	3-10	3
	3-2 × 8	5-3	2	4-0	2	3-5	2	5-1	2	4-0	2	3-5	2	4-7	2	3-11	2	3-4	2
ROOF, CEILING AND TWO FLOORS (CLEAR SPAN)	3-2 × 10	6-3	2	4-9	2	4-0	3	6-1	2	4-9	2	4-0	3	5-11	3	4-8	2	4-0	3
	3-2 × 12	7-5	3	5-8	3	4-9	3	7-2	3	5-8	3	4-8	3	5-6	3	5-6	2	4-8	2
	4-2 × 8	6-1	2	4-8	2	3-11	2	5-11	2	4-8	2	3-11	2	4-8	2	4-7	2	3-10	2
	4-2 × 10	7-3	2	5-6	2	4-8	2	7-0	2	5-6	2	4-7	2	5-5	2	5-5	2	4-7	2
	4-2 × 12	8-6	2	6-6	2	5-6	3	8-3	2	6-6	2	5-4	2	6-4	3	6-4	2	5-4	3

For SI: 1 inch = 25.4 mm, 1 pound per square foot = 0.0479 kPa.

a. Spans are given in feet and inches.
b. Spans are based on minimum design properties for No. 2 grade lumber of Douglas fir-larch, hem-fir, Southern pine, and spruce-pine-fir.
c. Building width is measured perpendicular to the ridge. For widths between those shown, spans are permitted to be interpolated.
d. NJ = Number of jack studs required to support each end. Where the number of required jack studs equals one, the header is permitted to be supported by an approved framing anchor attached to the full-height wall stud and to the header.
e. Use 30 psf ground snow load for cases in which ground snow load is less than 30 psf and the roof live load is equal to or less than 20 psf.
f. Spans are calculated assuming the top of the header or girder is laterally braced by perpendicular framing. Where the top of the header or girder is not laterally braced (for example, cripple studs bearing on the header), tabulated spans for headers consisting of 2 × 8, 2 × 10, or 2 × 12 sizes shall be multiplied by 0.70 or the header or girder shall be designed.

© 2018 IRC®, International Code Council®

HEADERS AND GIRDERS *(cont.)*

TABLE R602.7(2)
GIRDER SPANS[a] AND HEADER SPANS[a] FOR INTERIOR BEARING WALLS
(MAXIMUM SPANS FOR DOUGLAS FIR-LARCH, HEM-FIR, SOUTHERN PINE AND SPRUCE-PINE-FIR[b] AND REQUIRED NUMBER OF JACK STUDS)

HEADERS AND GIRDERS SUP-PORTING	SIZE	BUILDING Width[c] (feet)					
		12		24		36	
		Span[e]	NJ[d]	Span[e]	NJ[d]	Span[e]	NJ[d]
One floor only	2-2 × 4	4-1	1	2-10	1	2-4	1
	2-2 × 6	6-1	1	4-4	1	3-6	1
	2-2 × 8	7-9	1	5-5	1	4-5	2
	2-2 × 10	9-2	1	6-6	2	5-3	2
	2-2 × 12	10-9	1	7-7	2	6-3	2
	3-2 × 8	9-8	1	6-10	1	5-7	1
	3-2 × 10	11-5	1	8-1	1	6-7	2
	3-2 × 12	13-6	1	9-6	2	7-9	2
	4-2 × 8	11-2	1	7-11	1	6-5	1
	4-2 × 10	13-3	1	9-4	1	7-8	1
	4-2 × 12	15-7	1	11-0	1	9-0	2
Two floors	2-2 × 4	2-7	1	1-11	1	1-7	1
	2-2 × 6	3-11	1	2-11	2	2-5	2
	2-2 × 8	5-0	1	3-8	2	3-1	2
	2-2 × 10	5-11	2	4-4	2	3-7	2
	2-2 × 12	6-11	2	5-2	2	4-3	3
	3-2 × 8	6-3	1	4-7	2	3-10	2
	3-2 × 10	7-5	1	5-6	2	4-6	2
	3-2 × 12	8-8	2	6-5	2	5-4	2
	4-2 × 8	7-2	1	5-4	1	4-5	2
	4-2 × 10	8-6	1	6-4	2	5-3	2
	4-2 × 12	10-1	1	7-5	2	6-2	2

For SI:1 inch = 25.4 mm, 1 foot = 304.8 mm.

a. Spans are given in feet and inches.

b. Spans are based on minimum design properties for No. 2 grade lumber of Douglas fir-larch, hem-fir, Southern pine, and spruce-pine-fir.

c. Building width is measured perpendicular to the ridge. For widths between those shown, spans are permitted to be interpolated.

d. NJ = Number of jack studs required to support each end. Where the number of required jack studs equals one, the header is permitted to be supported by an approved framing anchor attached to the full-height wall stud and to the header.

e. Spans are calculated assuming the top of the header or girder is laterally braced by perpendicular framing. Where the top of the header or girder is not laterally braced (for example, cripple studs bearing on the header), tabulated spans for headers consisting of 2 × 8, 2 × 10, or 2 × 12 sizes shall be multiplied by 0.70 or the header or girder shall be designed.

HEADERS AND GIRDERS (cont.)

TABLE R602.7(3)
GIRDER AND HEADER SPANS[a] FOR OPEN PORCHES
(MAXIMUM SPAN FOR DOUGLAS FIR-LARCH, HEM-FIR, SOUTHERN PINE AND SPRUCE-PINE-FIR[b])

SIZE	SUPPORTING ROOF						SUPPORTING FLOOR	
	GROUND SNOW LOAD (psf)							
	30		50		70			
	DEPTH OF PORCH[c] (feet)							
	8	14	8	14	8	14	8	14
2-2 × 6	7-6	5-8	6-2	4-8	5-4	4-0	6-4	4-9
2-2 × 8	10-1	7-7	8-3	6-2	7-1	5-4	8-5	6-4
2-2 × 10	12-4	9-4	10-1	7-7	8-9	6-7	10-4	7-9
2-2 × 12	14-4	10-10	11-8	8-10	10-1	7-8	11-11	9-0

© 2018 INTERNATIONAL RESIDENTIAL CODE®

For SI:1 inch = 25.4 mm, 1 foot = 304.8 mm, 1 pound per square foot = 0.0479 kPa.

a. Spans are given in feet and inches.
b. Tabulated values assume No. 2 grade lumber, wet service and incising for refractory species. Use 30 psf ground snow load for cases in which ground snow load is less than 30 psf and the roof live load is equal to or less than 20 psf.
c. Porch depth is measured horizontally from building face to centerline of the header. For depths between those shown, spans are permitted to be interpolated.

Code	Description
R602.7	Header spans must be according to Tables R602.7(1) and (2) for exterior and interior bearing walls.
R602.7.4	Load-bearing headers are not required in interior or exterior nonbearing walls. A single, flat, 2″ by 4″ member may be used for openings up to 8′ in width if the vertical distance to the parallel nailing surface above is not more than 24″. For open porches, this table specifies a reduced size of girders and headers than that in Table R602.7(3).

You Should Know

- Wood structural panel box headers are allowed, and design criteria are depicted in Figure R602.7.3 in the IRC code book.
- Single-member headers are allowed per Section R602.7.1 in the IRC code book.
- Rim board headers have specific criteria specified in revised Section R602.7.2 of the IRC code book.

FIREBLOCKING

Where fireblocking is required:

1. In concealed spaces of stud walls and partitions, including furred spaces and parallel rows of studs or staggered studs, as follows:
   1.1. Vertically at the ceiling and floor levels.
   1.2. Horizontally at intervals not exceeding 10'.
2. At all interconnections between concealed vertical and horizontal spaces, such as at soffits, drop ceilings, and cove ceilings.
3. In concealed spaces between stair stringers at the top and bottom of the run. Enclosed spaces under stairs shall comply with Section R302.7 in the IRC code book.
4. At openings around vents, pipes, ducts, cables, and wires at ceiling and floor level, with an *approved* material to resist the free passage of flame and products of combustion. The material filling this annular space shall not be required to meet the ASTM E 136 requirements.
5. For the fireblocking of chimneys and fireplaces, see Section R1003.19 in the IRC code book.
6. Fireblocking of cornices of a two-family *dwelling* is required at the line of *dwelling unit* separation.

Types of fireblocking permitted:

1. Two-inch nominal lumber.
2. Two thicknesses of 1" nominal lumber with broken lap joints.
3. One thickness of $^{23}/_{32}$" wood structural panels with joints backed by $^{23}/_{32}$" wood structural panels.
4. One thickness of $^{3}/_{4}$" particleboard with joints backed by $^{3}/_{4}$" particleboard.
5. One-half-inch gypsum board.
6. One-quarter-inch, cement-based millboard.
7. Batts or blankets of mineral wool or glass fiber or other *approved* materials installed in such a manner as to be securely retained in place.
8. Cellulose insulation installed as tested in accordance with ASTM E119 or UL 263, for specific application.

Code	Description
R302.11	In combustible construction, fireblocking shall be provided to cut off all concealed draft openings (both vertical and horizontal) and to form an effective fire barrier between stories and between the top story and the roof space.
	Fireblocking shall be provided in wood frame construction in the locations indicated in the table above.
R302.11.1	Generally, the materials permitted for use as fireblocking are those depicted in the table above.

You Should Know

- Fireblocking is not required in buildings with concrete, masonry, or steel-framed walls.

DRAFT-STOPPING

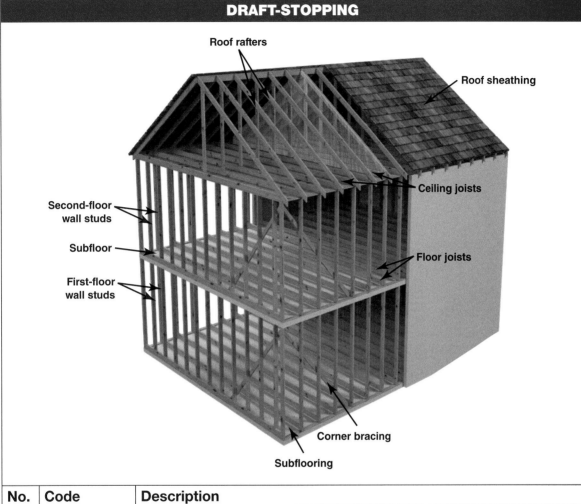

Roof rafters

Roof sheathing

Ceiling joists

Second-floor wall studs

Subfloor

Floor joists

First-floor wall studs

Corner bracing

Subflooring

No.	Code	Description
	R302.12	In combustible construction where there is usable space both above and below the concealed space of a floor/ceiling assembly, draft stops shall be installed so that the area of the concealed space does not exceed 1000 ft².
	R302.12.1	Draft-stopping materials shall not be less than ½" gypsum board, ⅜" wood structural panels, or other approved materials adequately supported.

You Should Know

- Fireblocking retards the hidden spread of fire within concealed elements of a frame assembly. Imagine a shaft that allows fire to rapidly spread from floor to floor. Stud wall cavities and similar openings have the same effect. Fireblocking limits this effect.

SIMPLIFIED WALL BRACING

TABLE R602.12.4 MINIMUM NUMBER OF BRACING UNITS ON EACH SIDE OF THE CIRCUMSCRIBED RECTANGLE

ULTIMATE DESIGN WIND SPEED (mph)	STORY LEVEL	EAVE-TO-RIDGE HEIGHT (feet)	MINIMUM NUMBER OF BRACING UNITS ON EACH LONG SIDE[a, b, d] Length of short side (feet)[c]						MINIMUM NUMBER OF BRACING UNITS ON EACH SHORT SIDE[a, b, d] Length of long side (feet)[c]					
			10	20	30	40	50	60	10	20	30	40	50	60
115	(3-story)	10	1	2	2	2	3	3	1	2	2	2	3	3
	(2-story)	10	2	3	3	4	5	6	2	3	3	4	5	6
	(1-story)	10	2	3	4	6	7	8	2	3	4	6	7	8
	(3-story)	15	1	2	3	3	4	4	1	2	3	3	4	4
	(2-story)	15	2	3	4	5	6	7	2	3	4	5	6	7
	(1-story)	15	2	4	5	6	7	9	2	4	5	6	7	9
130	(3-story)	10	1	2	2	3	3	4	1	2	2	3	3	4
	(2-story)	10	2	3	4	5	6	7	2	3	4	5	6	7
	(1-story)	10	2	4	5	7	8	10	2	4	5	7	8	10
	(3-story)	15	2	3	3	4	4	6	2	3	3	4	4	6
	(2-story)	15	3	4	6	7	8	10	3	4	6	7	8	10
	(1-story)	15	3	6	7	10	11	13	3	6	7	10	11	13

For SI: 1 inch = 25.4 mm, 1 foot = 304.8 mm.

a. Interpolation shall not be permitted.
b. Cripple walls or wood-framed basement walls in a walk-out condition shall be designated as the first story and the stories above shall be redesignated as the second and third stories, respectively, and shall be prohibited in a three-story structure.
c. Actual lengths of the sides of the circumscribed rectangle shall be rounded to the next highest unit of 10 when using this table.
d. For Exposure Category C, multiply bracing units by a factor of 1.20 for a one-story building, 1.30 for a two-story building and 1.40 for a three-story building.

©2018 IRC®, International Code Council®

WALL BRACING

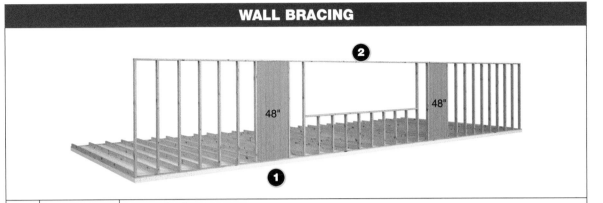

No.	Code	Description
1	R602.10	Buildings shall be braced according to this section or with an engineered design or, where applicable, Section R602.12 in the IRC code book ("Simplified Wall Bracing").
2	R602.10.1.1	The length of a *braced wall line* shall be the distance between its ends. The end of a *braced wall line* shall be the intersection with a perpendicular *braced wall line*, an angled *braced wall line* or an exterior wall.
	R602.10.4	The construction of intermittent braced wall panels shall be in accordance with one of the methods listed in Table R602.10.4 in the IRC code book. These are depicted in the illustration and are abbreviated as: • LIB: Let in bracing • DWB: Diagonal wall bracing • WSP: Wood structural panel • BV-WSP: Wood structural panel with stone or masonry veneer • SFB: Structural fiberboard sheathing • GB: Gypsum board • PBS: Particleboard sheathing • PCP: Portland cement plaster • HPS: Hardboard panel siding • ABW: Alternate brace wall panel • PFH: Intermittent portal frame • PFG: Intermittent portal frame at garage • CS-WSP: Continuously sheathed wood structural panel • CS-G: Continuously sheathed wood structural panel adjacent to garage • CS-PF: Continuously sheathed portal frame • CS-SFB: Continuously sheathed structural fiberboard
	R602.12	• Simplified wall bracing is now permitted for one to three story dwellings and town homes in wind exposure category B or C with ultimate design wind speeds of 130 mph or less.

You Should Know

- References above prescriptively meet the most common conditions. For areas affected by seismic and wind conditions, there are other design options. These are best addressed by a registered design professional.
- Section R602.10.4.4 requires panel joints to occur over framing members.

STEEL FRAMING

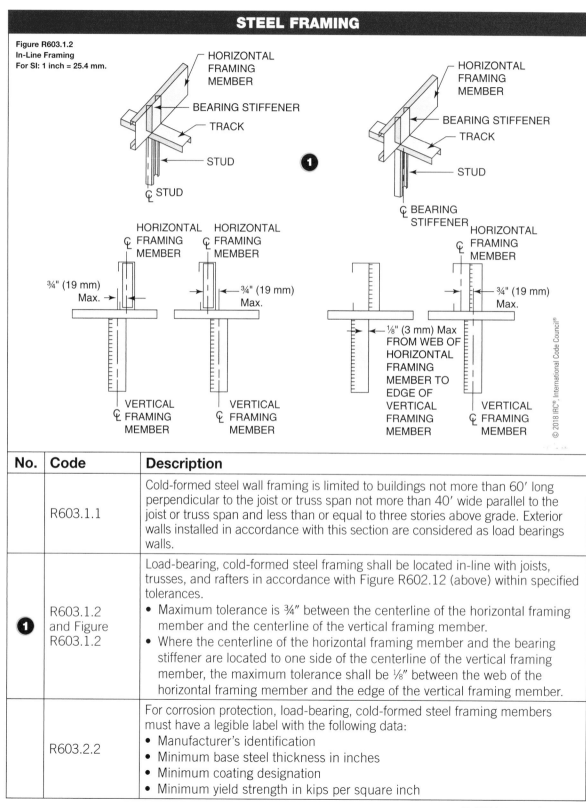

Figure R603.1.2
In-Line Framing
For SI: 1 inch = 25.4 mm.

No.	Code	Description
	R603.1.1	Cold-formed steel wall framing is limited to buildings not more than 60' long perpendicular to the joist or truss span not more than 40' wide parallel to the joist or truss span and less than or equal to three stories above grade. Exterior walls installed in accordance with this section are considered as load bearings walls.
❶	R603.1.2 and Figure R603.1.2	Load-bearing, cold-formed steel framing shall be located in-line with joists, trusses, and rafters in accordance with Figure R602.12 (above) within specified tolerances. • Maximum tolerance is ¾" between the centerline of the horizontal framing member and the centerline of the vertical framing member. • Where the centerline of the horizontal framing member and the bearing stiffener are located to one side of the centerline of the vertical framing member, the maximum tolerance shall be ⅛" between the web of the horizontal framing member and the edge of the vertical framing member.
	R603.2.2	For corrosion protection, load-bearing, cold-formed steel framing members must have a legible label with the following data: • Manufacturer's identification • Minimum base steel thickness in inches • Minimum coating designation • Minimum yield strength in kips per square inch

MASONRY WALLS

TABLE R606.9 ALLOWABLE COMPRESSIVE STRESSES FOR EMPIRICAL DESIGN OF MASONRY

CONSTRUCTION; COMPRESSIVE STRENGTH OF UNIT, GROSS AREA	ALLOWABLE COMPRESSIVE STRESSES[a] GROSS CROSS-SECTIONAL AREA[b]	
	Type M or S mortar	Type N mortar
Solid masonry of brick and other solid units of clay or shale; sand-lime or concrete brick:		
8,000 + psi	350	300
4,500 psi	225	200
2,500 psi	160	140
1,500 psi	115	100
Grouted[c] masonry, of clay or shale; sand-lime or concrete:		
4,500 + psi	225	200
2,500 psi	160	140
1,500 psi	115	100
Solid masonry of solid concrete masonry units:		
3,000 + psi	225	200
2,000 psi	160	140
1,200 psi	115	100
Masonry of hollow load-bearing units:		
2,000 + psi	140	120
1,500 psi	115	100
1,000 psi	75	70
700 psi	60	55
Hollow walls (cavity or masonry bonded[d]) solid units:		
2,500 + psi	160	140
1,500 psi	115	100
Hollow units	75	70
Stone ashlar masonry:		
Granite	720	640
Limestone or marble	450	400
Sandstone or cast stone	360	320
Rubble stone masonry:		
Coarse, rough or random	120	100

For SI: 1 pound per square inch = 6.895 kPa.

a. Linear interpolation shall be used for determining allowable stresses for masonry units having compressive strengths that are intermediate between those given in the table.
b. Gross cross-sectional area shall be calculated on the actual rather than nominal dimensions.
c. See Section R606.13.
d. Where floor and roof loads are carried upon one wythe, the gross cross-sectional area is that of the wythe under load; if both wythes are loaded, the gross cross-sectional area is that of the wall minus the area of the cavity between the wythes. Walls bonded with metal ties shall be considered as cavity walls unless the collar joints are filled with mortar or grout.

© 2018 IRC®, International Code Council®

TABLE R606.6.4 SPACING OF LATERAL SUPPORT FOR MASONRY WALLS

CONSTRUCTION	MAXIMUM WALL LENGTH TO THICKNESS OR WALL HEIGHT TO THICKNESS[a, b]
Bearing walls:	
Solid or solid grouted	20
All other	18
Nonbearing walls:	
Exterior	18
Interior	36

For SI: 1 foot = 304.8 mm.

a. Except for cavity walls and cantilevered walls, the thickness of a wall shall be its nominal thickness measured perpendicular to the face of the wall. For cavity walls, the thickness shall be determined as the sum of the nominal thicknesses of the individual wythes. For cantilever walls, except for parapets, the ratio of height to nominal thickness shall not exceed 6 for solid masonry, or 4 for hollow masonry. For parapets, see Section R606.4.4.
b. An additional unsupported height of 6 feet is permitted for gable end walls.

© 2018 IRC®, International Code Council®

MASONRY WALLS (cont.)

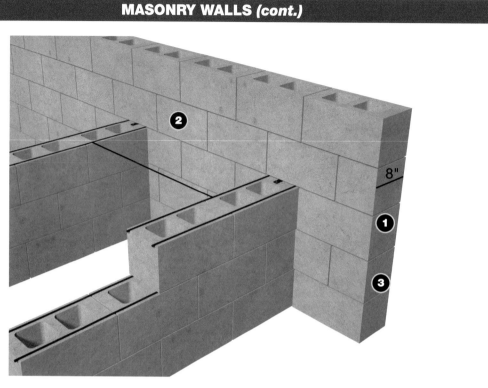

No.	Code	Description
1	R606.4.1	The minimum thickness of masonry bearing walls more than one story shall be 8". Solid masonry walls of one-story dwellings or garages shall be at least 6" in thickness and not more than 9' in height.
	R606.9	Allowable compressive stresses in masonry shall not exceed the values in Table R606.9 (above). Compressive stress is determined through product testing.
2	R606.7	The unsupported height of masonry piers shall not exceed 10 times their least dimension. If filled solidly with concrete or type M or S mortar. The unsupported height of masonry piers with unfilled hollow piers shall not exceed 4 times their least dimension.
	R606.6.4	Masonry walls shall be laterally supported in either the horizontal or vertical direction. The maximum spacing between lateral supports shall not exceed the distances in Table R606.4.
3	R606.4.1	The minimum thickness of masonry bearing walls more than one story high shall be 8".

You Should Know

- There is a minimum mortar cover of ⅝" for wall ties for aboveground masonry, Section R606.3.3 Item #1 in the IRC code book.
- Mortar head and bed joints must be ⅜" but have some tolerance: The bed joint may be ±⅛"; the head joint may be ±¼".

CONCRETE WALLS (INCLUDING INSULATED CONCRETE FORMS)

TABLE R608.3 DIMENSIONAL REQUIREMENTS FOR WALL[a, b]

WALL TYPE AND NOMINAL THICKNESS	MAXIMUM WALL WEIGHT[c] (psf)	MINIMUM WIDTH, W, OF VERTI-CAL CORES (inches)	MINIMUM THICKNESS, T, OF VERTI-CAL CORES (inches)	MAXIMUM SPACING OF VERTICAL CORES (inches)	MAXIMUM SPACING OF HORIZON-TAL CORES (inches)	MINIMUM WEB THICKNESS (inches)
4″ Flat[d]	50	N/A	N/A	N/A	N/A	N/A
6″ Flat[d]	75	N/A	N/A	N/A	N/A	N/A
8″ Flat[d]	100	N/A	N/A	N/A	N/A	N/A
10″ Flat[d]	125	N/A	N/A	N/A	N/A	N/A
6″ Waffle-grid	56	8[e]	5.5[e]	12	16	2
8″ Waffle-grid	76	8[f]	8[f]	12	16	2
6″ Screen-grid	53	6.25[g]	6.25[g]	12	12	N/A

For SI: 1 inch = 25.4 mm; 1 pound per square foot = 0.0479 kPa, 1 pound per cubic foot = 2402.77 kg/m³, 1 square inch = 645.16 mm², 1 inch⁴ = 42 cm⁴.

a. Width "W," thickness "T," spacing and web thickness, refer to Figures R608.3(2) and R608.3(3).

b. N/A indicates not applicable.

c. Wall weight is based on a unit weight of concrete of 150 pcf. For flat walls the weight is based on the nominal thickness. The tabulated values do not include any allowance for interior and exterior finishes.

d. Nominal wall thickness. The actual as-built thickness of a flat wall shall not be more than ½ inch less or more than ¼ inch more than the nominal dimension indicated.

e. Vertical core is assumed to be elliptical-shaped. Another shape core is permitted provided the minimum thickness is 5 inches, the moment of inertia, *I*, about the centerline of the wall (ignoring the web) is not less than 65 inch⁴, and the area, *A*, is not less than 31.25 square inches. The width used to calculate *A* and *I* shall not exceed 8 inches.

f. Vertical core is assumed to be circular. Another shape core is permitted provided the minimum thickness is 7 inches, the moment of inertia, *I*, about the centerline of the wall (ignoring the web) is not less than 200 inch⁴, and the area, *A*, is not less than 49 square inches. The width used to calculate *A* and *I* shall not exceed 8 inches.

g. Vertical core is assumed to be circular. Another shape core is permitted provided the minimum thickness is 5.5 inches, the moment of inertia, *I*, about the centerline of the wall is not less than 76 inch⁴, and the area, *A*, is not less than 30.25 square inches. The width used to calculate *A* and *I* shall not exceed 6.25 inches.

Figure R608.3(1)
Flat Wall System

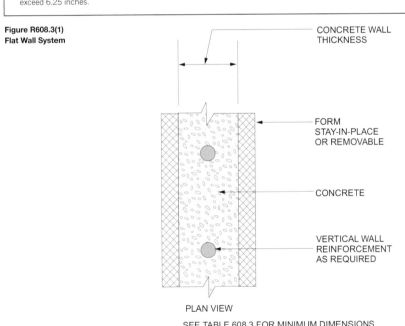

CONCRETE WALL THICKNESS

FORM STAY-IN-PLACE OR REMOVABLE

CONCRETE

VERTICAL WALL REINFORCEMENT AS REQUIRED

PLAN VIEW
SEE TABLE 608.3 FOR MINIMUM DIMENSIONS

CONCRETE WALLS (INCLUDING INSULATED CONCRETE FORMS) *(cont.)*

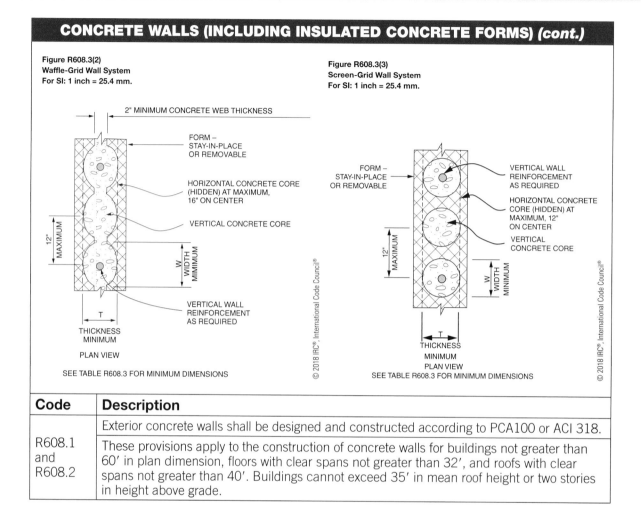

Figure R608.3(2)
Waffle-Grid Wall System
For SI: 1 inch = 25.4 mm.

Figure R608.3(3)
Screen-Grid Wall System
For SI: 1 inch = 25.4 mm.

Code	Description
R608.1 and R608.2	Exterior concrete walls shall be designed and constructed according to PCA100 or ACI 318.
	These provisions apply to the construction of concrete walls for buildings not greater than 60' in plan dimension, floors with clear spans not greater than 32', and roofs with clear spans not greater than 40'. Buildings cannot exceed 35' in mean roof height or two stories in height above grade.

You Should Know

- This section also applies to insulated concrete form (ICF) construction. These are known as *stay-in-place forms*. Prescriptive requirements for ICFs are based on the type of ICF, including flat grid, screen grid, and waffle grid systems as illustrated above.

- Reinforcement for concrete walls is determined based on design loads. Steel size and placement is similar to that found in foundation wall tables.

- Concrete's strength is primarily compressive. That is, it can support massive loads without crushing. Its compressive stress must be at least 2500 psi or pressure equal to 2500 pounds for every square inch of surface area. Core samples are sometimes taken and tested on the devices illustrated above.

MIXING CONCRETE

- Concrete is a mixture of cement, sand, aggregate gravel, and water. The strength of each mixture may vary based on the proportions of each ingredient. Durability and workability is also adjusted as designed based on the mixture formula. Sometimes admixtures are added to achieve these and other design conditions. Common admixtures include:
  - Retarding admixtures: slows the curing time of concrete
  - Accelerating admixtures: reduces the curing time for concrete
  - Super plasticizers: improves the workability of concrete
  - Water-reducing admixtures: maintains the workability while reducing the water content
  - Air-entraining admixtures: introduces air into mixture improving workability

TYPES OF CEMENT

I	Normal, general-purpose cement suitable for all normal uses
IA	Similar to Type I cement with air-entraining properties
II	Generates less heat at a slower rate
IIA	Similar to Type II with air-entraining properties
III	High early strength concrete that sets and gains strength rapidly
IIIA	Similar to Type III with air-entraining properties
IV	Develops strength at a slower rate and has a lower heat of hydration
V	Used only where concrete will be exposed to soil and groundwater with high sulfate content

STRUCTURAL INSULATED PANELS (SIPs)

Figure R610.5(1)
Maximum Allowable Height of SIP Walls
For SI: 1 foot = 304.8 mm.

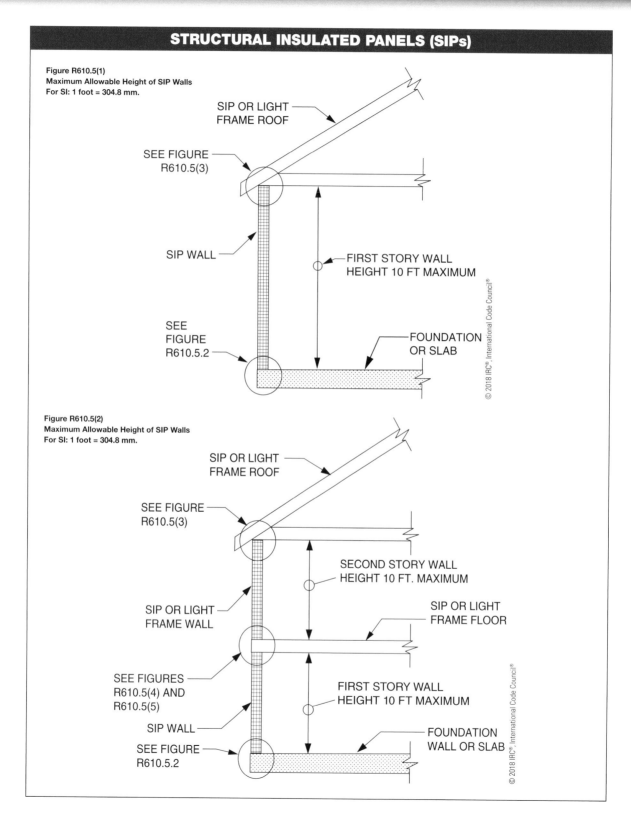

SIP OR LIGHT FRAME ROOF

SEE FIGURE R610.5(3)

SIP WALL

SEE FIGURE R610.5.2

FIRST STORY WALL HEIGHT 10 FT MAXIMUM

FOUNDATION OR SLAB

© 2018 IRC®, International Code Council®

Figure R610.5(2)
Maximum Allowable Height of SIP Walls
For SI: 1 foot = 304.8 mm.

SIP OR LIGHT FRAME ROOF

SEE FIGURE R610.5(3)

SIP OR LIGHT FRAME WALL

SEE FIGURES R610.5(4) AND R610.5(5)

SIP WALL

SEE FIGURE R610.5.2

SECOND STORY WALL HEIGHT 10 FT. MAXIMUM

SIP OR LIGHT FRAME FLOOR

FIRST STORY WALL HEIGHT 10 FT MAXIMUM

FOUNDATION WALL OR SLAB

© 2018 IRC®, International Code Council®

STRUCTURAL INSULATED PANELS (SIPs) *(cont.)*

Figure R610.5(4)
SIP Wall-to-Wall Platform Frame Connection
For SI: 1 inch = 25.4 mm.

Note: Figures illustrate SIP-specific attachment requirements. Other connections shall be made in accordance with Tables R602.3(1) and (2) as appropriate.

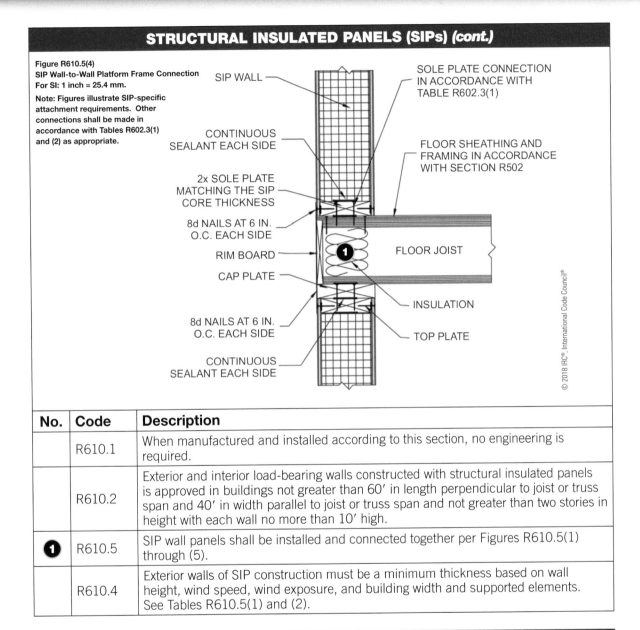

No.	Code	Description
	R610.1	When manufactured and installed according to this section, no engineering is required.
	R610.2	Exterior and interior load-bearing walls constructed with structural insulated panels is approved in buildings not greater than 60' in length perpendicular to joist or truss span and 40' in width parallel to joist or truss span and not greater than two stories in height with each wall no more than 10' high.
1	R610.5	SIP wall panels shall be installed and connected together per Figures R610.5(1) through (5).
	R610.4	Exterior walls of SIP construction must be a minimum thickness based on wall height, wind speed, wind exposure, and building width and supported elements. See Tables R610.5(1) and (2).

You Should Know

- SIPs are limited to areas with design wind speed no more than 155 mph Exposure B or 140 mph Exposure C or a ground snow load exceeding 70 lb/ft² and the seismic design category identification of A, B, or C.

PLASTER

TABLE R702.1(1) THICKNESS OF PLASTER

① PLASTER BASE	FINISHED THICKNESS OF PLASTER FROM FACE OF LATH, MASONRY, CONCRETE (inches)	
	Gypsum Plaster	Cement Plaster
Expanded metal lath	$5/8$, minimum[a]	$5/8$, minimum[a]
Wire lath	$5/8$, minimum[a]	$3/4$, minimum (interior)[b] $7/8$, minimum (exterior)[b]
Gypsum lath[g]	$1/2$, minimum	$3/4$, minimum (interior)[b]
Masonry walls[c]	$1/2$, minimum	$1/2$, minimum
Monolithic concrete walls[c, d]	$5/8$, maximum	$7/8$, maximum
Monolithic concrete ceilings[c, d]	$3/8$, maximum[e]	$1/2$, maximum
Gypsum veneer base[f, g]	$1/16$, minimum	$3/4$, minimum (interior)[b]
Gypsum sheathing[g]	—	$3/4$, minimum (interior)[b] $7/8$, minimum (exterior)[b]

For SI: 1 inch = 25.4 mm.

a. When measured from back plane of expanded metal lath, exclusive of ribs, or self-furring lath, plaster thickness shall be ¾ inch minimum.
b. When measured from face of support or backing.
c. Because masonry and concrete surfaces may vary in plane, thickness of plaster need not be uniform.
d. When applied over a liquid bonding agent, finish coat may be applied directly to concrete surface.
e. Approved acoustical plaster may be applied directly to concrete or over base coat plaster, beyond the maximum plaster thickness shown.
f. Attachment shall be in accordance with Table R702.3.5.
g. Where gypsum board is used as a base for cement plaster, a water-resistive barrier complying with Section R703.2 shall be provided.

© 2018 IRC®, International Code Council®

TABLE R702.1(2) GYPSUM PLASTER PROPORTIONS[A]

NUMBER	② COAT	PLASTER BASE OR LATH	MAXIMUM VOLUME AGGREGATE PER 100 POUNDS NEAT PLASTER[b] (cubic feet)	
			Damp Loose Sand[a]	Perlite or Vermiculite[c]
Two-coat work	Base coat	Gypsum lath	2.5	2
	Base coat	Masonry	3	3
Three-coat work	First coat	Lath	2[d]	2
	Second coat	Lath	3[d]	2[e]
	First and second coats	Masonry	3	3

For SI: 1 inch = 25.4 mm, 1 cubic foot = 0.0283 m³, 1 pound = 0.454 kg.

a. Wood-fibered gypsum plaster may be mixed in the proportions of 100 pounds of gypsum to not more than 1 cubic foot of sand where applied on masonry or concrete.
b. When determining the amount of aggregate in set plaster, a tolerance of 10 percent shall be allowed.
c. Combinations of sand and lightweight aggregate may be used, provided the volume and weight relationship of the combined aggregate to gypsum plaster is maintained.
d. If used for both first and second coats, the volume of aggregate may be 2.5 cubic feet.
e. Where plaster is 1 inch or more in total thickness, the proportions for the second coat may be increased to 3 cubic feet.

© 2018 IRC®, International Code Council®

PLASTER *(cont.)*

No.	Code	Description
❶	R702.1	Plaster must have a minimum thickness depending upon the plaster base and the type of plaster. See Table R702.1(1).
❷	Tables R702.1(2) and (3)	There are two major types of plaster: gypsum plaster and cement plaster. Each has specific proportions of ingredients. For gypsum plaster, sand, vermiculite, or perlite may be added. The volume of aggregate per 100 pounds of neat plaster varies from 2 to 3 ft^3.
	New Table R702.1(3)	Cement plaster is most often mixed on-site. There are normally three coats for this plaster. This table specifies the appropriate amounts based on the sequence of coats and other variables.

You Should Know

- There are four common types of gypsum plasters. Gypsum neat plaster is plain gypsum plaster without any additives such as aggregate. These are mixed on the jobsite. Gypsum ready-mixed plaster is gypsum mixed with aggregate. Gypsum wood-fibered plaster consists of calcined gypsum combined with wood fibers. Additives may be added to this mixture. Gypsum bond plaster is designed to bond to concrete.

- Portland cement plaster is similar to masonry mortar and normally site-mixed. Depending upon the first, second, or third coat, the proportions of ingredients vary.

GYPSUM BOARD

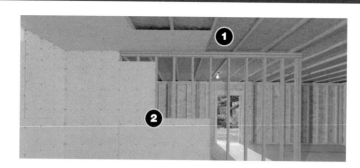

③ TABLE R702.3.5 MINIMUM THICKNESS AND APPLICATION OF GYPSUM BOARD AND GYPSUM PANEL PRODUCTS

THICKNESS OF GYPSUM BOARD OR GYPSUM PANEL PRODUCTS (inches)	APPLI-CATION	ORIENTATION OF GYPSUM BOARD OR GYPSUM PANEL PRODUCTS TO FRAMING	MAXIMUM SPACING OF FRAMING MEMBERS (inches o.c.)	MAXIMUM SPACING OF FASTENERS (inches)		SIZE OF NAILS FOR APPLICATION TO WOOD FRAMING[c]
				Nails[a]	Screws[b]	
Application without adhesive						
³⁄₈	Ceiling[d]	Perpendicular	16	7	12	13 gage, 1¼″ long, ¹⁹⁄₆₄″ head; 0.098″ diameter, 1¼″ long, annular-ringed; or 4d cooler nail, 0.080″ diameter, 1⅜″ long, ⁷⁄₃₂″ head.
³⁄₈	Wall	Either direction	16	8	16	
½	Ceiling	Either direction	16	7	12	13 gage, 1⅜″ long, ¹⁹⁄₆₄″ head; 0.098″ diameter, 1¼″ long, annular-ringed; 5d cooler nail, 0.086″ diameter, 1⅝″ long, ¹⁵⁄₆₄″ head; or gypsum board nail, 0.086″ diameter, 1⅝″ long, ⁹⁄₃₂″ head.
½	Ceiling[d]	Perpendicular	24	7	12	
½	Wall	Either direction	24	8	12	
½	Wall	Either direction	16	8	16	
⅝	Ceiling	Either direction	16	7	12	13 gage, 1⅝″ long, ¹⁹⁄₆₄″ head; 0.098″ diameter, 1⅜″ long, annular-ringed; 6d cooler nail, 0.092″ diameter, 1⅞″ long, ¼″ head; or gypsum board nail, 0.0915″ diameter, 1⅞″ long, ¹⁹⁄₆₄″ head.
⅝	Ceiling	Perpendicular	24	7	12	
⅝	Type X at garage ceiling beneath habitable rooms	Perpendicular	24	6	6	1⅞″ long 6d coated nails or equivalent drywall screws. Screws shall comply with Section R702.3.5.1
⅝	Wall	Either direction	24	8	12	13 gage, 1⅝″ long, ¹⁹⁄₆₄″ head; 0.098″ diameter, 1⅜″ long, annular-ringed; 6d cooler nail, 0.092″ diameter, 1⅞″ long, ¼″ head; or gypsum board nail, 0.0915″ diameter, 1⅞″ long, ¹⁹⁄₆₄″ head.
⅝	Wall	Either direction	16	8	16	
Application with adhesive						
³⁄₈	Ceiling[d]	Perpendicular	16	16	16	Same as above for ³⁄₈″ gypsum board and gypsum panel products.
³⁄₈	Wall	Either direction	16	16	24	
½ or ⅝	Ceiling	Either direction	16	16	16	Same as above for ½″ and ⅝″ gypsum board and gypsum panel products, respectively.
½ or ⅝	Ceiling[d]	Perpendicular	24	12	16	
½ or ⅝	Wall	Either direction	24	16	24	
Two ³⁄₈ layers	Ceiling	Perpendicular	16	16	16	Base ply nailed as above for ½″ gypsum board and gypsum panel products; face ply installed with adhesive.
Two ³⁄₈ layers	Wall	Either direction	24	24	24	

© 2018 IRC®, International Code Council®

(continues)

GYPSUM BOARD *(cont.)*

TABLE R702.3.5 *(cont.)*

For SI: 1 inch = 25.4 mm.

a. For application without adhesive, a pair of nails spaced not less than 2 inches apart or more than 2½ inches apart shall be permitted to be used with the pair of nails spaced 12 inches on center.

b. Screws shall be in accordance with Section R702.3.6. Screws for attaching gypsum board or gypsum panel products to structural insulated panels shall penetrate the wood structural panel facing not less than $7/16$ inch.

c. Where cold-formed steel framing is used with a clinching design to receive nails by two edges of metal, the nails shall be not less than ⅝ inch longer than the gypsum board or gypsum panel product thickness and shall have ringed shanks. Where the cold-formed steel framing has a nailing groove formed to receive the nails, the nails shall have barbed shanks or be 5d. 13½ gage, 1⅝ inches long, ¹⁵⁄₆₄-inch head for ½-inch gypsum board or gypsum panel product; and 6d, 13 gage, 1⅞ inches long, ¹⁵⁄₆₄-inch head for ⅝-inch gypsum board or gypsum panel product.

d. Three-eighths-inch-thick single-ply gypsum board or gypsum panel product shall not be used on a ceiling where a water-based textured finish is lo be applied, or where it will be required to support insulation above a ceiling. On ceiling applications to receive a water-based texture material, cither hand or spray applied, the gypsum board or gypsum panel product shall be applied perpendicular to framing. Where applying a water-based texture material, the minimum gypsum board thickness shall be increased from ⅜ inch to ½ inch for 16-inch on center framing, and from ½ inch to ⅝ inch for 24-inch on center framing or ½-inch sag-resistant gypsum ceiling board shall be used.

No.	Code	Description
❶	R701.2	Products sensitive to adverse weather shall not be installed until adequate weather protection for the installation is provided. Drywall may not be installed until the outside is dried in.
❷	R702.3.2	Wood framing support gypsum boards and gypsum panel products shall not be less than 2" nominal thickness in the least dimension except that 1" by 2" furring strips not less than 1" by 2" may be used over solid backing.
❸	R702.3.5	Maximum spacing of supports and the size and spacing of fasteners used to attach gypsum boards shall comply with Table R702.3.5.
	R702.3.1.1	Expandable foam adhesives for the installation of gypsum board and gypsum panels must conform to ASTM C6464.

You Should Know

- Do not install gypsum board until the final framing inspection has been approved according to Section R109.1.4 in the IRC code book.
- Gypsum board shall be applied perpendicular to ceiling framing members where used to create a horizontal diaphragm.

EXTERIOR WALL COVERING

Figure R703.8
Typical Masonry Veneer Wall Details
For SI: 1 inch = 34.5mm.

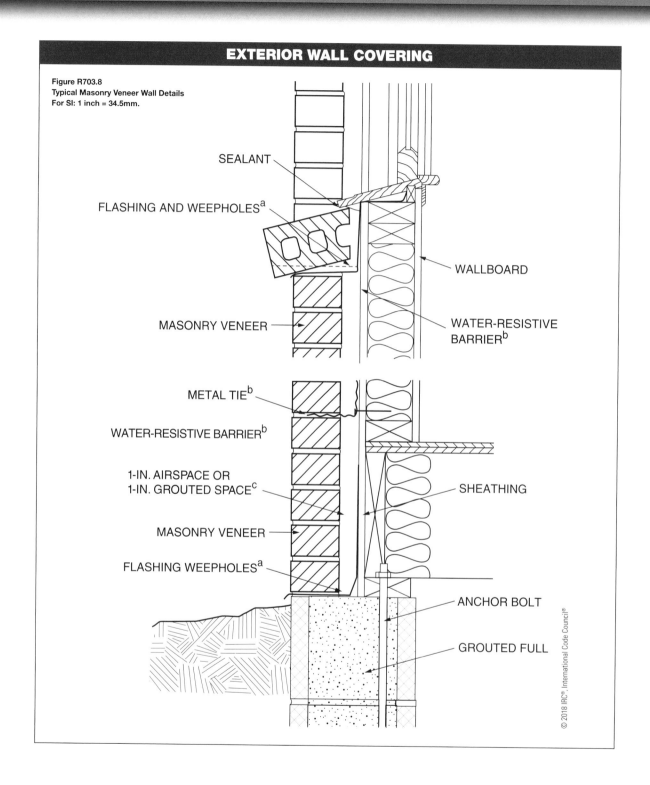

SEALANT

FLASHING AND WEEPHOLES[a]

WALLBOARD

MASONRY VENEER

WATER-RESISTIVE BARRIER[b]

METAL TIE[b]

WATER-RESISTIVE BARRIER[b]

1-IN. AIRSPACE OR 1-IN. GROUTED SPACE[c]

SHEATHING

MASONRY VENEER

FLASHING WEEPHOLES[a]

ANCHOR BOLT

GROUTED FULL

© 2018 IRC® International Code Council®

EXTERIOR WALL COVERING *(cont.)*

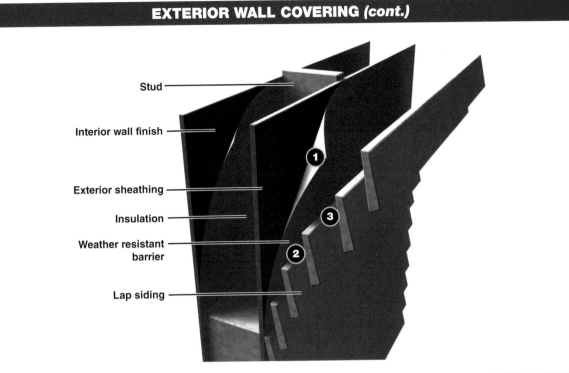

- Stud
- Interior wall finish
- Exterior sheathing
- Insulation
- Weather resistant barrier
- Lap siding

No.	Code	Description
1	R703.1.1	The exterior wall envelope shall be designed and constructed in a manner that prevents the accumulation of water within the wall assembly by providing a water-resistant barrier behind the exterior cladding and a means of draining to the exterior water that penetrates exterior cladding.
2	R703.2	One layer of #15 asphalt felt free from holes or breaks, complying with ASTM D226 for Type I felt, or other approved water-resistive barrier shall be applied over studs or sheathing of all exterior walls.
3		Such #15 asphalt felt shall be applied horizontally, with the upper layer lapped over the lower not less than 2″. Where joints occur, felt shall be lapped not less than 6″.
	R703.3	Fasteners for exterior wall coverings attached to wood framing shall be in accordance with Section R703.3 and Table R703.3(1) of the IRC.
	R703.3.4	Fasteners for hardboard panel and lap siding shall penetrate not less than 1½ inches into framing.

You Should Know

- Unless specified otherwise, all wall coverings shall be securely fastened in accordance with Table R703.4 in the IRC code book with approved aluminum, stainless-steel, zinc-coated, or other approved corrosion-resistant fasteners.

TABLE R703.3(1) SIDING MINIMUM ATTACHMENT AND MINIMUM THICKNESS

WEATHER-RESISTANT SIDING ATTACHMENT AND MINIMUM THICKNESS

SIDING MATERIAL	NOMINAL THICKNESS (inches)	JOINT TREATMENT	Wood or wood structural panel sheathing into stud	Fiberboard sheathing into stud	Gypsum sheathing into stud	Foam plastic sheathing into stud[l]	Direct to studs	Number or spacing of fasteners
Anchored veneer: brick, concrete, masonry or stone (see Section R703.8)	2	Section R703.8	Section R703.8					
Adhered veneer: concrete, stone or masonry (see Section R703.12)	—	Section R703.12	Section R703.12					
Fiber cement siding — Panel siding (see Section R703.10.1)	5/16	Section R703.10.1	6d common (2"×0.113")	6d common (2"×0.113")	6d common (2"×0.113")	6d common (2"×0.113")	4d common (1½"×0.099")	6" panel edges 12" inter. sup.
Fiber cement siding — Lap siding (see Section R703.10.2)	5/16	Section R703.10.2	6d common (2"×0.113")	6d common (2"×0.113")	6d common (2"×0.113")	6d common (2"×0.113")	6d common (2"×0.113") or 11 gage roofing nail	Note f
Hardboard panel siding (see Section R703.3)	7/16	—	0.120" nail (shank) with 0.225" head	0.120" nail (shank) with 0.225" head	0.120" nail (shank) with 0.225" head	0.120" nail (shank) with 0.225" head	0.120" nail (shank) with 0.225" head	6" panel edges 12" inter. sup.[d]
Hardboard lap siding (see Section R703.3)	7/16	Note e	0.099" nail (shank) with 0.240" head	0.099" nail (shank) with 0.240" head	0.099" nail (shank) with 0.240" head	0.099" nail (shank) with 0.240" head	0.099" nail (shank) with 0.240" head	Same as stud spacing 2 per bearing
Horizontal aluminum[a] — Without insulation	0.019[b]	Lap	Siding nail 1½"×0.120"	Siding nail 2"×0.120"	Siding nail 2"×0.120"	Siding nail[b] 1½"×0.120"	Not allowed	Same as stud spacing
Horizontal aluminum[a] — With insulation	0.024	Lap	Siding nail 1½"×0.120"	Siding nail 2"×0.120"	Siding nail 2"×0.120"	Siding nail[b] 1½"×0.120"	Not allowed	Same as stud spacing
	0.019	Lap	Siding nail 1½"×0.120"	Siding nail 2½"×0.120"	Siding nail 2½"×0.120"	Siding nail[b] 1½"×0.120"	Siding nail 1½"×0.120"	Same as stud spacing
Insulated vinyl siding[j]	0.035 (vinyl siding layer only)	Lap	0.120 nail (shank) with a 0.313 head or 16-gage crown[h,i]	0.120 nail (shank) with a 0.313 head or 16-gage crown[h]	0.120 nail (shank) with a 0.313 head or 16-gage crown[h]	0.120 nail (shank) with a 0.313 head Section R703.11.2	Not allowed	16 inches on center or specified by manufacturer instructions, test report or other sections of this code
Particleboard panels	3/8	—	6d box nail (2"×0.099")	6d box nail (2"×0.099")	6d box nail (2"×0.099")	6d box nail (2"×0.099")	Not allowed	6" panel edges, 12" inter. sup.
Particleboard panels	1/2	—	6d box nail (2"×0.099")	6d box nail (2"×0.099")	6d box nail (2"×0.099")	6d box nail (2"×0.099")	6d box nail (2"×0.099")	6" panel edges, 12" inter. sup.
Particleboard panels	5/8	—	6d box nail (2"×0.099")	8d box nail (2½"×0.113")	8d box nail (2½"×0.113")	6d box nail (2"×0.099")	6d box nail (2"×0.099")	6" panel edges, 12" inter. sup.
Polypropylene siding[k]	Not applicable	Lap	Section 703.14.1	Section 703.14.1	Section 703.14.1	Section 703.14.1	Not allowed	As specified by the manufacturer instructions, test report or other sections of this code

TYPE OF SUPPORTS FOR THE SIDING MATERIAL AND FASTENERS

(continues)

WEATHER-RESISTANT SIDING ATTACHMENT *(cont.)*

TABLE R703.3(1) *(cont.)*

SIDING MATERIAL		NOMINAL THICKNESS (inches)	JOINT TREATMENT	TYPE OF SUPPORTS FOR THE SIDING MATERIAL AND FASTENERS: Wood or wood structural panel sheathing into stud	Fiberboard sheathing into stud	Gypsum sheathing into stud	Foam plastic sheathing into stud[l]	Direct to studs	Number or spacing of fasteners
Steel[c]		29 ga.	Lap	Siding nail (1¾″ × 0.113″) Staple–1¾″	Siding nail (2¾″ × 0.113″) Staple–2½″	Siding nail (2½″ × 0.113″) Staple–2¼″	Siding nail (1¾″ × 0.113″) Staple–1¾″	Not allowed	Same as stud spacing
Vinyl siding (see Section R703.11)		0.035	Lap	0.120″ nail (shank) with a 0.313″ head or 16-gage staple with ⅜ - to ½-inch crown[h,i]	0.120″ nail (shank) with a 0.313″ head or 16-gage staple with ⅜-to ½-inch crown[h]	0.120″ nail (shank) with a 0.313″ head or 16- gage staple with ⅜ - to ½-inch crown[h]	0.120″ nail (shank) with a 0.313 head Section R703.11.2	Not allowed	16 inches on center or as specified by the manufacturer instructions or test report
Wood siding (see Section R703.3)	Wood rustic drop	⅜ min.	Lap	6d box or siding nail (2″ × 0.099″)	6d box or siding nail (2″ × 0.099″)	6d box or siding nail (2″ × 0.099″)	6d box or siding nail (2″ × 0.099″)	8d box or siding nail (2½″ × 0.113″) Staple–2″	Face nailing up to 6″ widths. 1 nail per bearing: 8″ widths and over, 2 nails per bearing
	Shiplap	19/32 average	Lap						
	Bevel	7/16							
	Bull lip	3/16	Lap						
Wood structural panel ANSI/APA PRP-210 siding (exterior grade) (see Section R703.3)		⅜ – ½	Note e	2″ × 0.099″ siding nail	2½″ × 0.113″ siding nail	2½″ × 0.113″ siding nail	2½″ × 0.113″ siding nail	2″ × 0.099″ siding nail	6″ panel edges 12″ inter. sup.
Wood structural panel lap siding (see Section R703.3)		⅜ – ½	Note e Note g	2″ × 0.099″ siding nail	2½″ × 0.113″ siding nail	2½″ × 0.113″ siding nail	2½″ × 0.113″ siding nail	2″ × 0.099″ siding nail	8″ along bottom edge

For SI: 1 inch = 25.4 mm.

a. Aluminum nails shall be used to attach aluminum siding.
b. Aluminum (0.019 inch) shall be unbacked only where the maximum panel width is 10 inches and the maximum flat area is 8 inches. The tolerance for aluminum siding shall be +0.002 inch of the nominal dimension.
c. Shall be of approved type.
d. Where used to resist shear forces, the spacing must be 4 inches at panel edges and 8 inches on interior supports.
e. Vertical end joints shall occur at studs and shall be covered with a joint cover or shall be caulked.
f. Face nailing: one 6d common nail through the overlapping planks at each stud. Concealed nailing: one 11-gage 1½-inch-long galv. roofing nail through the top edge of each plank at each stud in accordance with the manufacturer's installation instructions.
g. Vertical joints, if staggered, shall be permitted to be away from studs if applied over wood structural panel sheathing.
h. Minimum fastener length must be sufficient to penetrate sheathing other nailable substrate and framing a total of a minimum of 1¼ inches or in accordance with the manufacturer's installation instructions.
i. Where specified by the manufacturer's instructions and supported by a test report, fasteners are permitted to penetrate into or fully through nailable sheathing or other nailable substrate of minimum thickness specified by the instructions or test report, without penetrating into framing.
j. Insulated vinyl siding shall comply with ASTM D 7793.
k. Polypropylene siding shall comply with ASTM D 7254.
l. Cladding attachment over foam sheathing shall comply with the additional requirements and limitations of Sections R703.15, R703.16 and R703.17.

MASONRY VENEER

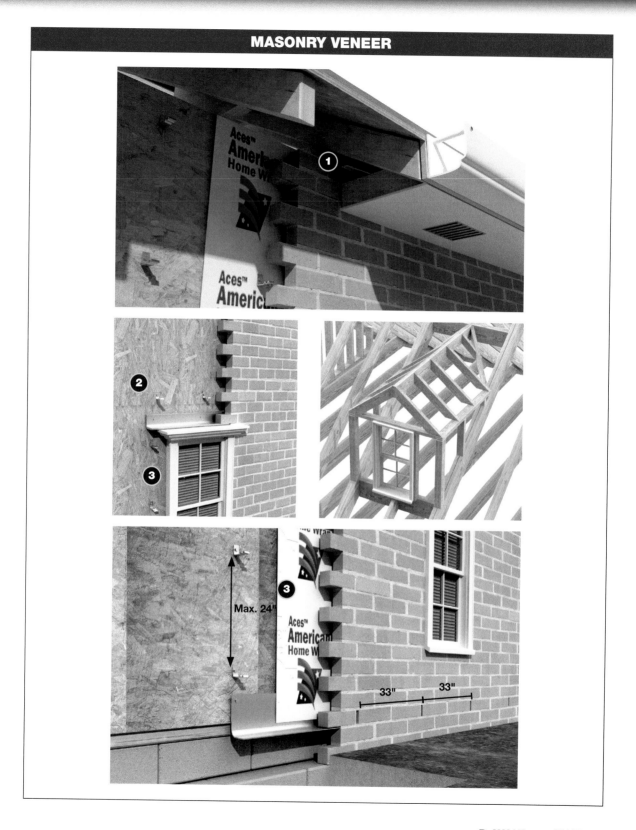

No.	Code	Description
1	R703.8	These veneers installed over the backing of wood or cold-formed steel shall be limited to the first story above grade plane and shall not exceed 5″ in thickness.
2	R703.8.4	Masonry veneer shall be anchored to the supporting wall with corrosion-resistant metal ties embedded in mortar or grout extending into the veneer a minimum of 1½″.
3	R703.8.4.1	Veneer ties shall be spaced not more than 32″ on center horizontally and 24″ on center vertically and each tie shall support no more than 2.67 ft^2 of wall area.

MASONRY VENEER *(cont.)*

You Should Know

- The above veneer information is for buildings in a lower seismic design category.

FLASHING

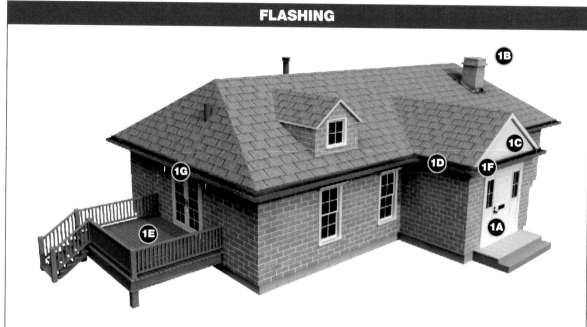

No.	Code	Description
1	R703.4	Approved corrosion-resistant flashing shall be applied shingle-fashion in the following locations: A. Exterior window and door openings B. Intersection of chimneys and other masonry construction C. Under and at the ends of masonry, wood, or metal copings and sill D. Continuously above all projecting wood trim E. Where exterior porches, decks, or stairs attach to the wall or floor assembly of wood frame construction F. At wall and roof intersections G. At built-in gutters
	R903.2	Flashing must be installed in a manner that prevents moisture from entering the wall and roof through joints in copings, through moisture-permeable materials, and at intersections with parapet walls and other penetrations through the roof plane.
	R903.2.1	Flashing must be installed at wall and roof intersections, whenever there is a change in roof slope or direction, and around roof openings.

You Should Know

- The methods of installing flashing are established by the manufacturer. Look for the manufacturer's installation instructions. See Section R703.4 in the IRC code book.

RIDGE BOARD FRAMING

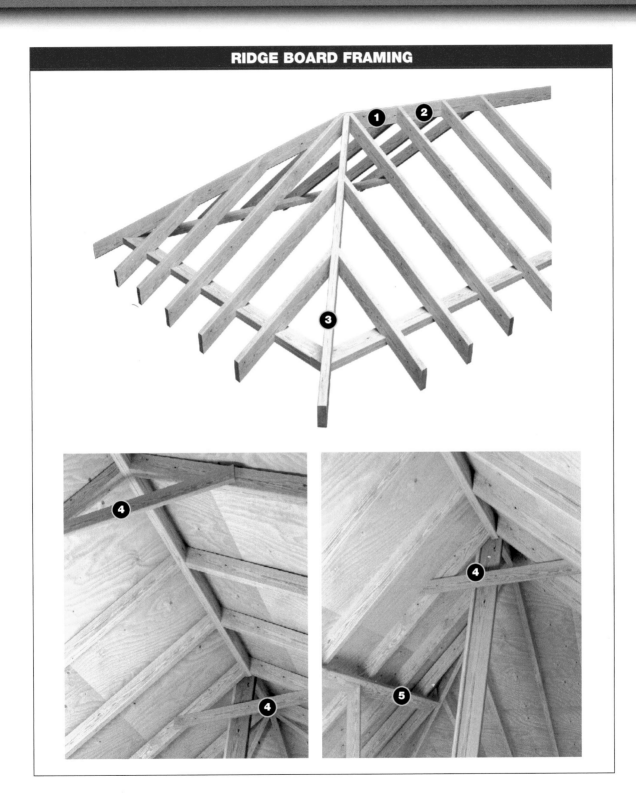

RIDGE BOARD FRAMING *(cont.)*

No.	Code	Description
❶		Rafters shall be framed to the ridge board or to each other with a gusset plate or continuous ties across the structure.
❷	R802.3	Ridge board shall be at least 1″ nominal thickness and not less than the cut end of the rafter.
❸		At all valleys and hips, there shall be a valley or hip rafter not less than 2″ nominal thickness and not less than the cut end of the rafter.
❹	R802.3, R802.4.3	Hip and valley rafters shall be supported at the ridge by a brace to a bearing partition or be designed to carry and distribute the specific load at that point. Where the roof pitch is less than 3 units vertical to 12 units horizontal, structural members that support rafters and ceiling joists such as a ridge beam, hips, and valleys shall be designed as a beam.
❺	R802.4.6	Collar ties shall be a minimum of 1″ by 4″ spaced not more than 4′ on center.
	Tables R802.4.1 (1) to (8)	Design values for Southern Pine lumber are incorporated into this table for ceiling joists (see following pages).

You Should Know

- Ridge board framing is a common framing method involving opposing rafters tied together at the base, forming a triangle.
- Ridge beam framing is another common framing method that uses a structural beam at the ridge that supports rafters from one or both sides.
- Section R802.3 in the IRC code book includes framing details for roof construction.
- Rafters must be framed to a ridge board (or to each other) with a gusset plate or a tie.
- Ends of ceiling joists must be lapped a minimum of 3 or butted over bearing partitions or beams and toe-nailed to the bearing member.
- All wood rafters and ceiling joists shall have not less than 1½ of bearing on wood or steel and 3 of bearing on masonry or concrete.

RIDGE BOARD FRAMING *(cont.)*

TABLE R802.4.1(1)
RAFTER SPANS FOR COMMON LUMBER SPECIES
(ROOF LIVE LOAD = 20 PSF, CEILING NOT ATTACHED TO RAFTERS, L/Δ = 180)

RAFTER SPACING (inches)	SPECIES AND GRADE		DEAD LOAD = 10 psf					DEAD LOAD = 20 psf				
			2 × 4	2 × 6	2 × 8	2 × 10	2 × 12	2 × 4	2 × 6	2 × 8	2 × 10	2 × 12
			Maximum rafter spans[a]									
			(feet-inches)	(feet-inches)	(feet-inches)	(feet-inches)	(feet-inches)	(feet-inches)	(feet-inches)	(feet-inches)	(feet-inches)	(feet-inches)
12	Douglas fir-larch	SS	11-6	18-0	23-9	Note b	Note b	11-6	18-0	23-9	Note b	Note b
	Douglas fir-larch	#1	11-1	17-4	22-5	Note b	Note b	10-6	15-4	19-5	23-9	Note b
	Douglas fir-larch	#2	10-10	16-10	21-4	26-0	Note b	10-0	14-7	18-5	22-6	26-0
	Douglas fir-larch	#3	8-9	12-10	16-3	19-10	23-0	7-7	11-1	14-1	17-2	19-11
	Hem-fir	SS	10-10	17-0	22-5	Note b	Note b	10-10	17-0	22-5	Note b	Note b
	Hem-fir	#1	10-7	16-8	22-0	Note b	Note b	10-4	15-2	19-2	23-5	Note b
	Hem-fir	#2	10-1	15-11	20-8	25-3	Note b	9-8	14-2	17-11	21-11	25-5
	Hem-fir	#3	8-7	12-6	15-10	19-5	22-6	7-5	10-10	13-9	16-9	19-6
	Southern pine	SS	11-3	17-8	23-4	Note b	Note b	11-3	17-8	23-4	Note b	Note b
	Southern pine	#1	10-10	17-0	22-5	Note b	Note b	10-6	15-8	19-10	23-2	Note b
	Southern pine	#2	10-4	15-7	19-8	23-5	Note b	9-0	13-6	17-1	20-3	23-10
	Southern pine	#3	8-0	11-9	14-10	18-0	21-4	6-11	10-2	12-10	15-7	18-6
	Spruce-pine-fir	SS	10-7	16-8	21-11	Note b	Note b	10-7	16-8	21-9	Note b	Note b
	Spruce-pine-fir	#1	10-4	16-3	21-0	25-8	Note b	9-10	14-4	18-2	22-3	25-9
	Spruce-pine-fir	#2	10-4	16-3	21-0	25-8	Note b	9-10	14-4	18-2	22-3	25-9
	Spruce-pine-fir	#3	8-7	12-6	15-10	19-5	22-6	7-5	10-10	13-9	16-9	19-6
16	Douglas fir-larch	SS	10-5	16-4	21-7	Note b	Note b	10-5	16-3	20-7	25-2	Note b
	Douglas fir-larch	#1	10-0	15-4	19-5	23-9	Note b	9-1	13-3	16-10	20-7	23-10
	Douglas fir-larch	#2	9-10	14-7	18-5	22-6	26-0	8-7	12-7	16-0	19-6	22-7
	Douglas fir-larch	#3	7-7	11-1	14-1	17-2	19-11	6-7	9-8	12-12	14-11	17-3
	Hem-fir	SS	9-10	15-6	20-5	Note b	Note b	9-10	15-6	19-11	24-4	Note b
	Hem-fir	#1	9-8	15-2	19-2	23-5	Note b	9-0	13-1	16-7	20-4	23-7
	Hem-fir	#2	9-2	14-2	17-11	21-11	25-5	8-5	12-3	15-6	18-11	22-0
	Hem-fir	#3	7-5	10-10	13-9	16-9	19-6	6-5	9-5	11-11	14-6	16-10
	Southern pine	SS	10-3	16-1	21-2	Note b	Note b	10-3	16-1	21-2	25-7	Note b
	Southern pine	#1	9-10	15-6	19-10	23-2	Note b	9-1	13-7	17-2	20-1	23-10
	Southern pine	#2	9-0	13-6	17-1	20-3	23-10	7-9	11-8	14-9	17-6	20-8
	Southern pine	#3	6-11	10-2	12-10	15-7	18-6	6-0	8-10	11-2	13-6	16-0
	Spruce-pine-fir	SS	9-8	15-2	19-11	25-5	Note b	9-8	14-10	18-10	23-0	Note b
	Spruce-pine-fir	#1	9-5	14-4	18-2	22-3	25-9	8-6	12-5	15-9	19-3	22-4
	Spruce-pine-fir	#2	9-5	14-4	18-2	22-3	25-9	8-6	12-5	15-9	19-3	22-4
	Spruce-pine-fir	#3	7-5	10-10	13-9	16-9	19-6	6-5	9-5	11-11	14-6	16-10
19.2	Douglas fir-larch	SS	9-10	15-5	20-4	25-11	Note b	9-10	14-10	18-10	23-0	Note b
	Douglas fir-larch	#1	9-5	14-0	17-9	21-8	25-2	8-4	12-2	15-4	18-9	21-9
	Douglas fir-larch	#2	9-1	13-3	16-10	20-7	23-10	7-10	11-6	14-7	17-10	20-8
	Douglas fir-larch	#3	6-11	10-2	12-10	15-8	18-3	6-0	8-9	11-2	12-7	15-9
	Hem-fir	SS	9-3	14-7	19-2	24-6	Note b	9-3	14-4	18-2	22-3	25-9
	Hem-fir	#1	9-1	13-10	17-6	21-5	24-10	8-2	12-0	15-2	18-6	21-6
	Hem-fir	#2	8-8	12-11	16-4	20-0	23-2	7-8	11-2	14-2	17-4	20-1
	Hem-fir	#3	6-9	9-11	12-7	15-4	17-9	5-10	8-7	10-10	13-3	15-5
	Southern pine	SS	9-8	15-2	19-11	25-5	Note b	9-8	15-2	19-7	23-4	Note b
	Southern pine	#1	9-3	14-3	18-1	21-2	25-2	8-4	12-4	15-8	18-4	21-9
	Southern pine	#2	8-2	12-3	15-7	18-6	21-9	7-1	10-8	13-6	16-0	18-10
	Southern pine	#3	6-4	9-4	11-9	14-3	16-10	5-6	8-1	10-2	12-4	14-7
	Spruce-pine-fir	SS	9-1	14-3	18-9	23-11	Note b	9-1	13-7	17-2	21-0	24-4
	Spruce-pine-fir	#1	8-10	13-1	16-7	20-3	23-6	7-9	11-4	14-4	17-7	20-4
	Spruce-pine-fir	#2	8-10	13-1	16-7	20-3	23-6	7-9	11-4	14-4	17-7	20-4
	Spruce-pine-fir	#3	6-9	9-11	12-7	15-4	17-9	5-10	8-7	10-10	13-3	15-5

(continues)

RIDGE BOARD FRAMING *(cont.)*

TABLE R802.4.1(1) *(cont.)*

RAFTER SPACING (inches)	SPECIES AND GRADE	DEAD LOAD = 10 psf					DEAD LOAD = 20 psf				
		2 × 4	2 × 6	2 × 8	2 × 10	2 × 12	2 × 4	2 × 6	2 × 8	2 × 10	2 × 12
		Maximum rafter spans[a]									
		(feet-inches)	(feet-inches)	(feet-inches)	(feet-inches)	(feet-inches)	(feet-inches)	(feet-inches)	(feet-inches)	(feet-inches)
24	Douglas fir-larch SS	9-1	14-4	18-10	23-9	Note b	9-1	13-3	16-10	20-7	23-10
	Douglas fir-larch #1	8-7	12-6	15-10	19-5	22-6	7-5	10-10	13-9	16-9	19-6
	Douglas fir-larch #2	8-2	11-11	15-1	18-5	21-4	7-0	10-4	13-0	15-11	18-6
	Douglasc fir-larch #3	6-2	9-1	11-6	14-1	16-3	5-4	7-10	10-0	12-2	14-1
	Hem-fir SS	8-7	13-6	17-10	22-9	Note b	8-7	12-10	16-3	19-10	23-0
	Hem-fir #1	8-5	12-4	15-8	19-2	22-2	7-4	10-9	13-7	16-7	19-3
	Hem-fir #2	7-11	11-7	14-8	17-10	20-9	6-10	10-0	12-8	15-6	17-11
	Hem-fir #3	6-1	8-10	11-3	13-8	15-11	5-3	7-8	9-9	11-10	13-9
	Southern pine SS	8-11	14-1	18-6	23-8	Note b	8-11	13-10	17-6	20-10	24-8
	Southern pine #1	8-7	12-9	16-2	18-11	22-6	7-5	11-1	14-0	16-5	19-6
	Southern pine #2	7-4	11-0	13-11	16-6	19-6	6-4	9-6	12-1	14-4	16-10
	Southern pine #3	5-8	8-4	10-6	12-9	15-1	4-11	7-3	9-1	11-0	13-1
	Spruce-pine-fir SS	8-5	13-3	17-5	21-8	25-2	8-4	12-2	15-4	18-9	21-9
	Spruce-pine-fir #1	8-0	11-9	14-10	18-2	21-0	6-11	10-2	12-10	15-8	18-3
	Spruce-pine-fir #2	8-0	11-9	14-10	18-2	21-0	6-11	10-2	12-10	15-8	18-3
	Spruce-pine-fir #3	6-1	8-10	11-3	13-8	15-11	5-3	7-8	9-9	11-10	13-9

Check sources for availability of lumber in lengths greater than 20 feet.

For SI: 1 inch = 25.4 mm, 1 foot = 304.8 mm, 1 pound per square foot = 0.0479 kPa.

a. The tabulated rafter spans assume that ceiling joists are located at the bottom of the attic space or that some other method of resisting the outward push of the rafters on the bearing walls, such as rafter ties, is provided at that location. Where ceiling joists or rafter ties are located higher in the attic space, the rafter spans shall be multiplied by the following factors:

H_C/H_R	Rafter Span Adjustment Factor
1/3	0.67
1/4	0.76
1/5	0.83
1/6	0.90
1/7.5 or less	1.00

where:

H_C = Height of ceiling joists or rafter ties measured vertically above the top of the rafter support walls.

H_R = Height of roof ridge measured vertically above the top of the rafter support walls.

b. Span exceeds 26 feet in length.

© 2018 IRC®, International Code Council®

CONSTRUCTION

WOOD TRUSSES

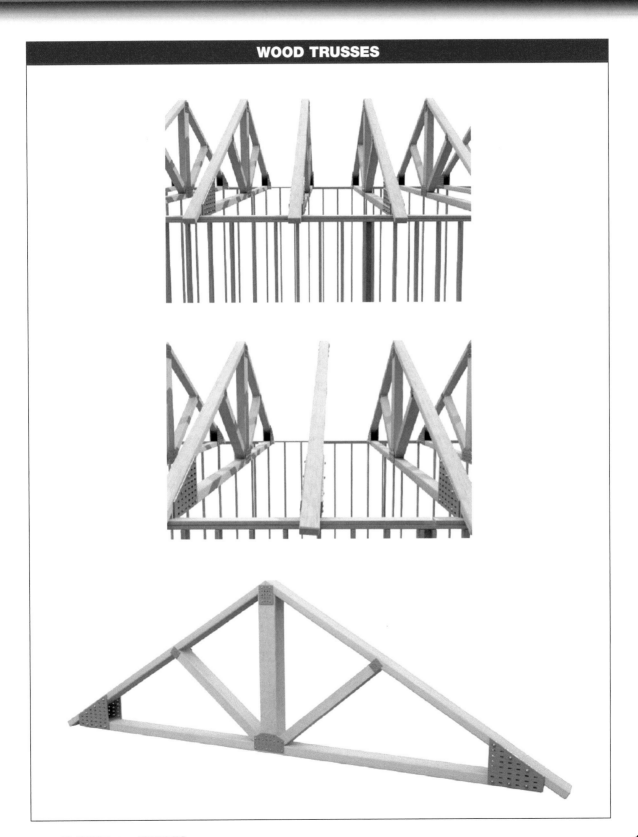

WOOD TRUSSES *(cont.)*

Code	Description
R802.10.1	Truss design drawings in conformance with Section R802.10.1 in the IRC code book shall be provided to the building official and approved prior to installation.
R802.10.2.1	Wood trusses are limited to buildings not greater than 60' in length perpendicular to the truss span, not greater than 36' in width parallel to the truss, not greater than two stories in height with each story not greater than 10' high, and roof slopes not smaller than 3:12 or greater than 12:12.
R802.10.3	Trusses shall be braced to prevent rotation and provide lateral stability.
R802.10.4	Truss members shall not be cut, notched, drilled, spliced, or otherwise altered in any way without approval from a registered design professional.
R802.11.1.1	Trusses shall be connected to wall plates by use of approved connectors having a resistance uplift as specified on the truss design drawings for the ultimate design wind speed.

You Should Know

- Wood trusses are engineered wood products. As such, they are designed and manufactured according to specifications based on conditions. The code lists information that must be provided in a truss drawing in order to be approved. These items include:

    1. Slope or depth, span, and spacing
    2. Location of all joints
    3. Required bearing widths
    4. Design loads
    5. Adjustments to lumber and joint connector design values
    6. Each reaction force and direction
    7. Joint connector type
    8. Lumber size, species, and grade for each member
    9. Connection requirements truss-girder, truss ply to ply, field splices
    10. Deflection ratio and maximum live and total load
    11. Maximum axial compression forces in truss members
    12. Required permanent truss member bracing locations

MANUFACTURED AND ENGINEERED TRUSSES

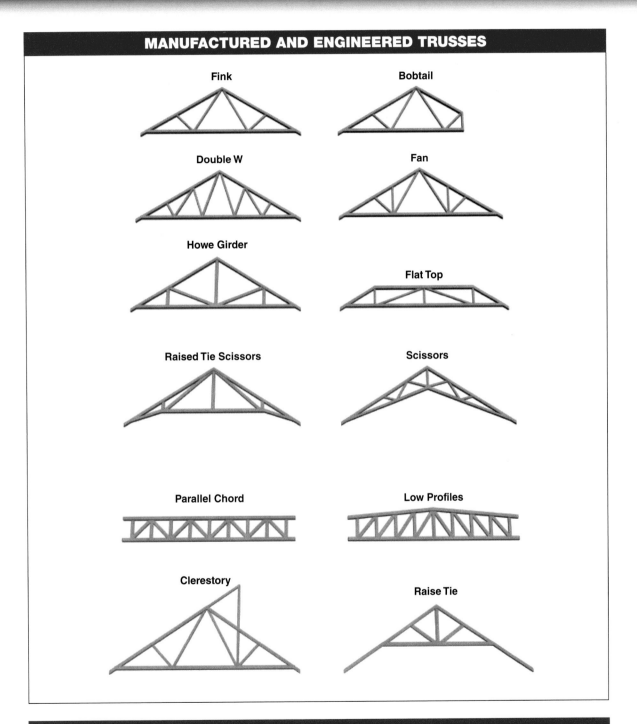

Fink

Bobtail

Double W

Fan

Howe Girder

Flat Top

Raised Tie Scissors

Scissors

Parallel Chord

Low Profiles

Clerestory

Raise Tie

You Should Know

- Spans for rafters shall be in accordance with Tables R802.4(1) through (8) in the IRC code book.
- Spans for ceiling joists shall be in accordance with Tables R802.5(1) and (2) in the IRC code book.

ROOF VENTILATION AND ATTIC ACCESS

No.	Code	Description
	R806.1 and R806.2	The minimum net free ventilating area shall be $\frac{1}{150}$ of the area of the vented space except: 1. In climate zones 6, 7, and 8 where a Class I or II vapor retarder is installed on the warm-in-winter side of the ceiling. 2. At least 40% and not more than 50% of required ventilation is provided in the upper portion of the attic or rafter space.

ROOF VENTILATION AND ATTIC ACCESS *(cont.)*

No.	Code	Description
	R806.5	Unvented attic assemblies and unvented enclosed rafter assemblies shall be permitted if several conditions are met: 1. The unvented attic space is completely contained within the building thermal envelope. 2. No interior Class I vapor retarders are installed on the ceiling side (attic floor) of the unvented attic assembly or on the ceiling side of the unvented enclosed rafter assembly. 3. Where wood shingles or shakes are used, a minimum ¼" (6 mm) vented air space separates the shingles or shakes and the roofing underlayment above the structural sheathing. 4. In climate zones 5, 6, 7, and 8, any air-impermeable insulation shall be a Class II vapor retarder or shall have a Class III vapor retarder coating or covering in direct contact with the underside of the insulation. 5. Either items 5.1, 5.2, or 5.3 shall be met, depending on the air permeability of the insulation directly under the structural roof sheathing. 5.1.1 Where only air-permeable insulation is installed directly below the structural roof sheathing. 5.1.2 Where air-permeable insulation is provided inside the building thermal envelope, it shall be installed in accordance with 5.1.1 above. In addition, rigid board or sheet insulation shall be installed above the structural roof sheathing with *R*-values as required by Table R806.5. 5.1.3 Where both air impermeable and air-permeable insulation are provided, the air impermeable shall be applied in direct contact with the underside of the structural roof sheathing. The air-permeable insulation shall be installed directly under the air-impermeable insulation. 5.1.4 Alternately, sufficient rigid board or sheet insulation shall be installed directly above the structural roof sheathing to maintain the monthly average temperature of the underside of the structural roof sheathing above 45° F. 5.2 Climate Zones 1, 2, and 3 installed in unvented attics shall follow 10 conditions set out in Section R806.5.
①	R807.1	Attic access openings are required in buildings with a combustible ceiling or roof if the attic has 30" or more of vertical height or greater over an area of not less than 30 sq. ft. The rough-framed opening must be at least 22" by 30" and in a hallway or other location with ready access. When located in a ceiling, there must be at least 30" of headroom in the attic space.

You Should Know

- Required ventilation openings shall open directly to the outside air.
- There are additional access requirements for mechanical equipment located in attics.

DRAINAGE

30"

Sheet metal cricket

1

Inlet flow 2"above low point of roof

4" min.

2

Inlet 2" above low point of roof

3

No.	Code	Description
DRAINAGE *(cont.)*		
❶	R903.2.2	A cricket or saddle shall be installed on the ridge side of any chimney or penetration more than 30″ wide as measured perpendicular to the slope.
❷	R903.4	Unless roofs are sloped to drain over roof edges, roof drains shall be installed at each low point of the roof.
❸		Where roof drains are required, secondary emergency overflow drains or scuppers shall be placed level with the perimeter construction extending above the roof where water could be trapped. The scupper shall be located as determined by the roof slope and contributing roof area.

You Should Know

- Roof flashing is required at wall and roof intersections, where there is a change in roof slope or direction, and around roof openings, as illustrated.
- Unit skylights flashed per manufacturer's instructions do not need a cricket or a saddle. See Section R903.2.2 exception in the IRC code book.

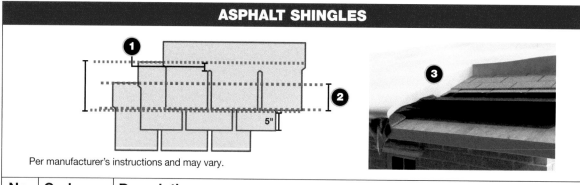

ASPHALT SHINGLES

Per manufacturer's instructions and may vary.

No.	Code	Description
1	R905.2.5	Fasteners for asphalt shingles shall be galvanized steel, stainless steel, aluminum, or copper roofing nails, minimum 12-gauge [0.105″ (3 mm)] shank with a minimum ⅜″ (10 mm) diameter head, ASTM F 1667, of a length to penetrate through the roofing materials and a minimum of ¾″ (19 mm) into the roof sheathing. Where the roof sheathing is less than ¾″ (19 mm) thick, the fasteners shall penetrate through the sheathing.
	R905.2.6	Asphalt shingles shall have the minimum number of fasteners required by the manufacturer's approved installation instructions, but not less than four fasteners per strip shingle or two fasteners per individual shingle.
2	R905.1.1	Underlayment for asphalt shingles must comply with ASTM D226, D1970, D4869, and D6757 and shall bear the label indicating compliance to the standard designation.
3	R905.1.2	In areas where there has been a history of ice forming along the eaves causing a backup of water as designated in Table R301.2(1) in the IRC code book, an ice barrier that consists of a least two layers of underlayment cemented together or of a self-adhering polymer modified bitumen sheet shall be used in lieu of normal underlayment and extend from the lowest edges of all roof surfaces to a point at least 24″ (610 mm) inside the exterior wall line of the building.

You Should Know

- A drip edge is required at eaves and gables on shingle roofs and fastened every 12″. See Section R905.2.8.5 in the IRC code book.

CLAY AND CONCRETE TILE

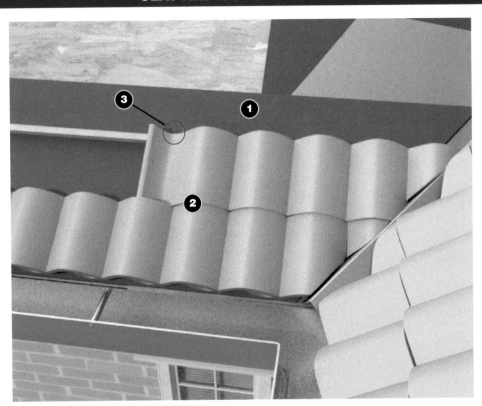

TABLE R905.3.7 CLAY AND CONCRETE TILE ATTACHMENT

SHEATHING	ROOF SLOPE	NUMBER OF FASTENERS
Solid without battens	All	One per tile
Spaced or solid with battens and slope < 5:12	Fasteners not required	—
Spaced sheathing without battens	5:12 ≤ slope < 12:12	One per tile/every other row
	12:12 ≤ slope < 24:12	One per tile

© 2018 IRC®, International Code Council®

No.	Code	Description
❶	R905.3.1	Concrete and clay tile shall be installed only over solid sheathing or spaced structural sheathing boards.
❷	R905.3.2	Clay and concrete roof tile shall be installed on roof slopes of 2½ units vertical in 12 units horizontal (2½:12) or greater. For roof slopes from 2½ units vertical in 12 units horizontal (2½:12) to 4 units vertical in 12 units horizontal (4:12), double underlayment application is required in accordance with Section R905.3.3 in the IRC code book.
❸	R905.3.6	Nails shall be corrosion-resistant and not less than 11 gauge, $\frac{5}{16}$" (11 mm) head, and of sufficient length to penetrate the deck a minimum of ¾" (19 mm) or through the thickness of the deck, whichever is less. Attaching wire for clay or concrete tile shall not be smaller than 0.083" (2 mm). Perimeter fastening areas include three tile courses but not less than 36" (914 mm) from either side of hips or ridges and edges of eaves and gable rakes.

WOOD SHINGLES

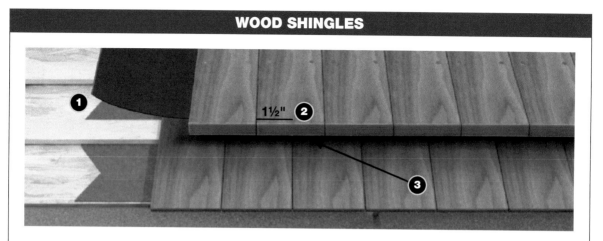

TABLE R905.7.5(1) WOOD SHINGLE WEATHER EXPOSURE AND ROOF SLOPE

ROOFING MATERIAL	LENGTH (inches)	GRADE	EXPOSURE (inches)	
			3:12 pitch to < 4:12	4:12 pitch or steeper
Shingles of naturally durable wood	16	No. 1	3¾	5
	16	No. 2	3½	4
	16	No. 3	3	3½
	18	No. 1	4¼	5½
	18	No. 2	4	4½
	18	No. 3	3½	4
	24	No. 1	5¾	7½
	24	No. 2	5½	6½
	24	No. 3	5	5½

For SI: 1 inch = 25.4 mm.

© 2018 IRC®, International Code Council®

TABLE R905.7.5(2) NAIL REQUIREMENTS FOR WOOD SHAKES AND WOOD SHINGLES

SHAKES	NAIL TYPE AND MINIMUM LENGTH	MINIMUM HEAD SIZE	MINIMUM SHANK DIAMETER
18″ straight-split	5d box 1¾″	0.19″	.080″
18″ and 24″ handsplit and resawn	6d box 2″	0.19″	.0915″
24″ taper-split	5d box 1¾″	0.19″	.080″
18″ and 24″ tapersawn	6d box 2″	0.19″	.0915″
Shingles	**Nail Type and Minimum Length**		
16″ and 18″	3d box 1¼″	0.19″	.080″
24″	4d box 1½″	0.19″	.080″

© 2018 IRC®, International Code Council®

WOOD SHINGLES *(cont.)*

No.	Code	Description
1	R905.7.1	Wood shingles shall be installed on solid or spaced sheathing. Where spaced sheathing is used, sheathing boards shall not be less than 1″ by 4″ (25.4 mm by 102 mm) nominal dimensions and shall be spaced on centers equal to the weather exposure to coincide with the placement of fasteners.
	R905.7.2	Wood shingles shall be installed on slopes of 3 units vertical in 12 units horizontal (25% slope) or greater.
	R905.7.3.1	Where required, ice barriers shall comply with Section R905.1.2 that states: In areas where there has been a history of ice forming along the eaves causing a backup of water, an ice barrier that consists of at least two layers of underlayment cemented together or a self-adhering polymer modified bitumen sheet shall be used in lieu of normal underlayment and extend from the lowest edges of all roof surfaces to a point at least 240″ (610 mm) inside the exterior wall line of the building.
2	R905.7.5	Wood shingles shall be laid with a side lap not less than 1½″ (38 mm) between joints in courses, and no two joints in any three adjacent courses shall be in direct alignment.
3		Spacing between shingles shall not be less than ¼″ to ⅜″ (6 mm to 10 mm).
4		Weather exposure for wood shingles shall not exceed those set in Table R905.7.5(1) (above).
		Fasteners for untreated wood shingles shall be box nails in accordance with Table R905.7.5(2) above. Nails shall be stainless steel type 304 or 316 or hot dipped galvanized, with a coating weight per ASTM A153 Class D. There are alternative fasteners but note that if within 15 miles of salt water coastal areas you must use stainless steel Type 316.

WOOD SHAKES

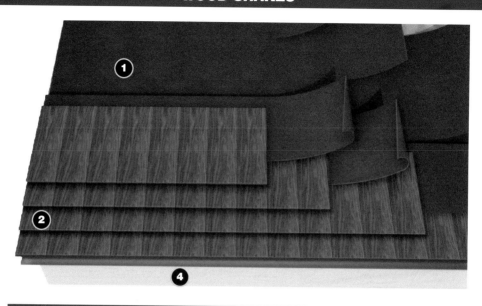

TABLE R905.8.6 WOOD SHAKE WEATHER EXPOSURE AND ROOF SLOPE

❸ ROOFING MATERIAL	LENGTH (inches)	GRADE	EXPOSURE (inches) 4:12 pitch or steeper
Shakes of naturally durable wood	18	No. 1	7½
	24	No. 1	10ª
Preservative-treated taper-sawn shakes of Southern Yellow Pine	18	No. 1	7½
	24	No. 1	10
	18	No. 2	5½
	24	No. 2	7½
Taper-sawn shakes of naturally durable wood	18	No. 1	7½
	24	No. 1	10
	18	No. 2	5½
	24	No. 2	7½

For SI: 1 inch = 25.4 mm.

a. For 24-inch by ⅜-inch handsplit shakes, the maximum exposure is 7½ inches.

© 2018 IRC®, International Code Council®

No.	Code	Description
❶	R905.8	Wood shakes shall be used only on solid or spaced sheathing. Where spaced sheathing is used, sheathing boards shall not be less than 1″ by 4″ (25 mm by 102 mm) nominal dimensions and shall be spaced on centers equal to the weather exposure to coincide with the placement of fasteners. Where 1″ by 4″ spaced sheathing is installed at 10″ (254 mm) on center, additional 1″ by 4″ boards shall be installed between the sheathing boards.
	R905.8.2	Wood shakes shall only be used on slopes of 3 units vertical in 12 units horizontal (25% slope) or greater.

No.	Code	Description
	R905.8.3.1	Where required, ice barriers shall comply with Section R905.1.2 that states: In areas where there has been a history of ice forming along the eaves causing a backup of water, an ice barrier that consists of at least two layers of underlayment cemented together or a self-adhering polymer modified bitumen sheet shall be used in lieu of normal underlayment and extend from the lowest edges of all roof surfaces to a point at least 240" (610 mm) inside the exterior wall line of the building.
❷	R905.8.6	Shakes shall be laid with a side lap not less than 1½" (38 mm) between joints in adjacent courses. Spacing between shakes in the same course shall be ⅜" to ⅝" (9.5 mm to 15.9 mm) for shakes and taper-sawn shakes of naturally durable wood and shall be ⅜" to ⅝" (9.5 mm to 15.9 mm) for taper-sawn shakes.
❸	R905.8.6	Weather exposure for wood shakes shall not exceed those set forth in Table R905.8.6.
	R905.8.6	Fasteners for untreated wood shakes shall be box nails in accordance with Table R905.7.5(2) above. Nails shall be stainless steel type 304 or 316 or hot dipped galvanized, with a coating weight per ASTM A153 Class D. There are alternative fasteners but note that if within 15 miles of salt water coastal areas you must use stainless steel Type 316.
❹	R905.8.7	The starter course at the eaves shall be doubled and the bottom layer shall be either 15" (381 mm), 18" (457 mm), or 24" (610 mm) wood shakes or wood shingles. Fifteen-inch (381 mm) or 18" (457 mm) wood shakes may be used for the final course at the ridge. Shakes shall be interlaid with 18"-wide (457 mm) strips of not less than No. 30 felt shingled between each course in such a manner that no felt is exposed to the weather by positioning the lower edge of each felt strip above the butt end of the shake it covers a distance equal to twice the weather exposure.

PHOTOVOLTAIC SHINGLES

Photo Courtesy of Beldon Roofing Company

No.	Code	Description
	R905.16	Installation of photovoltaic shingles shall comply with this section, R324 and NFPA 70 including: 905.16.1 Deck requirements, Deck slope, Underlayment, Underlayment application and Ice Barrier.
	R905.17	Building-integrated photovoltaic (BIPV) roof panels applied directly to the roof deck must comply with Section R324 and NFPA 70.

MASONRY FIREPLACES

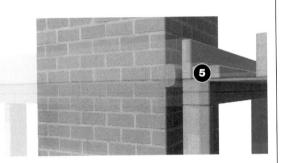

No.	Code	Description
	R1001.2 and M1801.3.3	Footings for masonry fireplaces and their chimneys shall be constructed of concrete or solid masonry at least 12″ (305 mm) thick and shall extend at least 6″ (152 mm) beyond the face of the fireplace or foundation wall on all sides. Footings shall be founded on natural, undisturbed earth or engineered fill below frost depth. Masonry chimneys shall be provided with a cleanout opening. Cleanout openings shall be provided within 6″ (152 mm) of the base of each flue within every masonry chimney. The upper edge of the cleanout shall be located at least 6″ below the lowest chimney inlet opening. The height of the opening shall be at least 6″. The cleanout shall be provided with a noncombustible cover.
❶	R1001.6	The firebox of a concrete or masonry fireplace shall have a minimum depth of 20″ (508 mm). The throat shall not be less than 8″ (203 mm) above the fireplace opening. The throat opening shall not be less than 4″ (102 mm) deep.
❷	R1001.9.1	The minimum thickness of fireplace hearths shall be 4″ (102 mm).
❸	R1001.9.2	The minimum thickness of hearth extensions shall be 2″ (51 mm). However, when the bottom of the firebox opening is raised at least 8″ (203 mm) above the top of the hearth extension, a hearth extension of not less than ⅜″-thick (10 mm) brick, concrete, stone, tile, or other approved noncombustible material is permitted.
❹	R1001.10	Hearth extensions shall extend at least 16″ (406 mm) in front of and at least 8″ (203 mm) beyond each side of the fireplace opening. Where the fireplace opening is 6 ft² (0.6 m²) or larger, the hearth extension shall extend at least 20″ (508 mm) in front of and at least 12″ (305 mm) beyond each side of the fireplace opening.
❺	R1001.11	All wood beams, joists, studs, and other combustible material shall have a clearance of not less than 2″ (51 mm) from the front faces and sides of masonry fireplaces and not less than 4″ (102 mm) from the back faces of masonry fireplaces.

You Should Know

- In areas not subjected to freezing, footings shall be at least 12″ (305 mm) below finished grade.
- Hearth extensions for factory-built fireplaces must be according to the listing of the fireplace.

MASONRY CHIMNEYS

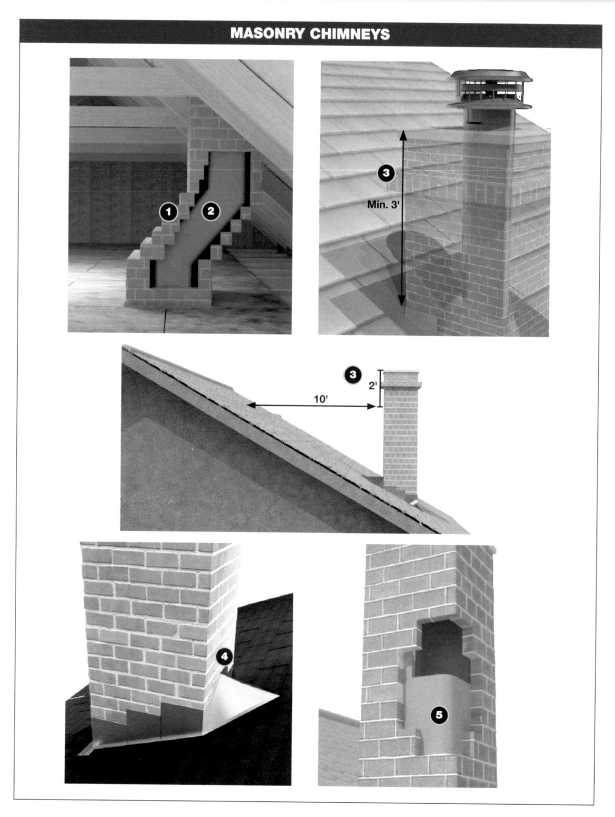

MASONRY CHIMNEYS (cont.)

No.	Code	Description
	R1003.2	Footings for masonry chimneys shall be constructed of concrete or solid masonry at least 12" (305 mm) thick and shall extend at least 6" (152 mm) beyond the face of the foundation or support wall on all sides.
1	R1003.5	Masonry chimneys shall not be corbelled more than one-half of the chimney's wall thickness from a wall or foundation, nor shall a chimney be corbelled from a wall or foundation that is less than 12" (305 mm) thick unless it projects equally on each side of the wall.
2	R1003.7	Where a masonry chimney is constructed with a fireclay flue liner surrounded by one wythe of masonry, the maximum offset shall be such that the centerline of the flue above the offset does not extend beyond the center of the chimney wall below the offset.
3	R1003.9	Chimneys shall extend at least 2' (610 mm) higher than any portion of a building within 10' (3048 mm) but shall not be less than 3' (914 mm) above the highest point where the chimney passes through the roof.
4	R1003.10	Masonry chimney walls shall be constructed of solid masonry units or hollow masonry units grouted solid with not less than a 4" (102 mm) nominal thickness.
5	R1003.11	Masonry chimneys shall be lined. The lining material shall be appropriate for the type of appliance connected, according to the terms of the appliance listing and manufacturer's instructions.

You Should Know

- Factory-built or masonry fireplaces must be equipped with an exterior air supply to assure proper combustion unless the room is mechanically ventilated and controlled so that indoor pressure is neutral or positive.
- Chimney caps are required per Section R1003.9.1 in the IRC code book. This section specifies that a concrete, metal, or stone cap, sloped to shed water, must be installed.

INSULATION AND FENESTRATION REQUIREMENTS BY COMPONENT

TABLE N1102.1.2 (R402.1.2) INSULATION AND FENESTRATION REQUIREMENTS BY COMPONENT[a]

CLIMATE ZONE	FENESTRATION U-FACTOR[b]	SKYLIGHT[b] U-FACTOR	GLAZED FENESTRATION SHGC[b, e]	CEILING R-VALUE	WOOD FRAME WALL R-VALUE	MASS WALL R-VALUE[i]	FLOOR R-VALUE	BASEMENT[c] WALL R-VALUE	SLAB[d] R-VALUE & DEPTH	CRAWL SPACE[c] WALL R-VALUE
1	NR	0.75	0.25	30	13	3/4	13	0	0	0
2	0.40	0.65	0.25	38	13	4/6	13	0	0	0
3	0.35	0.55	0.25	38	20 or 13 + 5[h]	8/13	19	5/13[f]	0	5/13
4 except Marine	0.35	0.55	0.40	49	20 or 13 + 5[h]	8/13	19	10/13	10, 2 ft	10/13
5 and Marine 4	0.32	0.55	NR	49	20 or 13 + 5[h]	13/17	30[g]	15/19	10, 2 ft	15/19
6	0.32	0.55	NR	49	20 + 5 or 13 + 10[h]	15/20	30[g]	15/19	10, 4 ft	15/19
7 and 8	0.32	0.55	NR	49	20 + 5 or 13 + 10[h]	19/21	38[g]	15/19	10, 4 ft	15/19

For SI: 1 foot = 304.8 mm.

a. R-values are minimums. U-factors and SHGC are maximums. When insulation is installed in a cavity which is less than the label or design thickness of the insulation, the installed R-value of the insulation shall not be less than the R-value specified in the table.

b. The fenestration U-factor column excludes skylights. The SHGC column applies to all glazed fenestration.
   Exception: Skylights may be excluded from glazed fenestration SHGC requirements in Climate Zones 1 through 3 where the SHGC for such skylights does not exceed 0.30.

c. "15/19" means R-15 continuous insulation on the interior or exterior of the home or R-19 cavity insulation at the interior of the basement wall, "15/19" shall be permitted to be met with R-13 cavity insulation on the interior of the basement wall plus R-5 continuous insulation on the interior or exterior of the home. "10/13" means R-10 continuous insulation on the interior or exterior of the home or R-13 cavity insulation at the interior of the basement wall.

d. R-5 shall be added to the required slab edge R-values for heated slabs. Insulation depth shall be the depth of the footing or 2 feet, whichever is less in Zones 1 through 3 for heated slabs.

e. There are no SHGC requirements in the Marine Zone.

f. Basement wall insulation is not required in warm-humid locations as defined by Figure N1101.10 and Table N1101.10.

g. Or insulation sufficient to fill the framing cavity, R-19 minimum.

h. The first value is cavity insulation, the second value is continuous insulation, so "13+5" means R-13 cavity insulation plus R-5 continuous insulation.

i. The second R-value applies when more than half the insulation is on the interior of the mass wall.

© 2018 IRC®, International Code Council®

Code	Description
Figure N1101.7 and Table N1101.7	Climate zones, moisture regimes, and warm-humid designations are set out by state, county, and territory in this table. The zones range from 1 to 7, with 1 being the warmest and 7 being the coldest.
N1102.1 and Table N1102.1.2	Compliance shall be demonstrated by meeting the requirements either of the International Energy Conservation Code or of N1102.1. Climate zones from Figure N1101.7 or Table N1101.7.2(2) in the IRC code book shall be used in determining the applicable requirements.

You Should Know

- R-value is the inverse of μ-value. Insulation is normally rated in R-value and windows are normally rated in μ-value. The μ-value decreases as the ability to resist heat flow increases, unlike R-value that increases as capacity to resist heat flow increases. Both are within the table that regulates insulation and fenestration requirements. Table N1102.1.2 in the IRC code book provides equivalent μ factors for those items noted as R-values in the table above.

- Fenestration or openings have two separate requirements in the table above. While there is a μ-value for all climate zones, there is a solar heat gain coefficient (SHGC) for climate zones 1–3.

CONSTRUCTION

AIR LEAKAGE

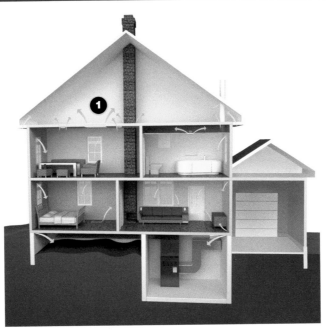

No.	Code	Description
	N1102.4	N1102.4.1 Building thermal envelope. The building thermal envelope shall comply with Sections N1102.4.1.1 and N1102.4.1.2 in the IRC code book and be constructed to limit air leakage.
1	N1103.3.2	Ducts, air handlers, and filter boxes used as ducts shall be sealed. Joints and seams shall comply with this code.

You Should Know

- Building envelope air tightness and insulation installation must comply with Section N1103.3.3 of the IRC code book.

BUILDING THERMAL ENVELOPE

TABLE N1102.4.1.1 (R402.4.1.1) AIR BARRIER AND INSULATION INSTALLATION[a]

COMPONENT	AIR BARRIER CRITERIA	INSULATION INSTALLATION CRITERIA
General requirements	A continuous air barrier shall be installed in the building envelope. The exterior thermal envelope contains a continuous air barrier. Breaks or joints in the air barrier shall be sealed.	Air-permeable insulation shall not be used as a sealing material.
Ceiling/attic	The air barrier in any dropped ceiling or soffit shall be aligned with the insulation and any gaps in the air barrier sealed. Access openings, drop down stairs or knee wall doors to unconditioned attic spaces shall be sealed.	The insulation in any dropped ceiling/soffit shall be aligned with the air barrier.
Walls	The junction of the foundation and sill plate shall be sealed. The junction of the top plate and the top of exterior walls shall be sealed. Knee walls shall be sealed.	Cavities within corners and headers of frame walls shall be insulated by completely filling the cavity with a material having a thermal resistance of not less than R-3 per inch. Exterior thermal envelope insulation for framed walls shall be installed in substantial contact and in continuous alignment with the air barrier.
Windows, skylights and doors	The space between framing and skylights, and the jambs of windows and doors, shall be sealed.	—
Rim joists	Rim joists shall include the air barrier.	Rim joists shall be insulated.
Floors including cantilevered floors and floors above garages.	The air barrier shall be installed at any exposed edge of insulation.	Floor framing cavity insulation shall be installed to maintain permanent contact with the underside of subfloor decking. Alternatively, floor framing cavity insulation shall be in contact with the top side of sheathing or continuous insulation installed on the underside of floor framing; and extending from the bottom to the top of all perimeter floor framing members.
Crawl space walls	Exposed earth in unvented crawl spaces shall be covered with a Class I vapor retarder with overlapping joints taped.	Crawl space insulation, where provided instead of floor insulation, shall be permanently attached to the walls.
Shafts, penetrations	Duct shafts, utility penetrations, and flue shafts opening to exterior or unconditioned space shall be sealed.	—
Narrow cavities	—	Batts to be installed in narrow cavities shall be cut to fit or narrow cavities shall be filled with insulation that on installation readily conforms to the available cavity space.
Garage separation	Air sealing shall be provided between the garage and conditioned spaces.	—
Recessed lighting	Recessed light fixtures installed in the building thermal envelope shall be sealed to the finished surface.	Recessed light fixtures installed in the building thermal envelope shall be air tight and IC rated.
Plumbing and wiring	—	In exterior walls, batt insulation shall be cut neatly to fit around wiring and plumbing or insulation that on installation, readily conforms to available space, shall extend behind piping and wiring.
Shower/tub on exterior wall	The air barrier installed at exterior walls adjacent to showers and tubs shall separate the wall from the shower or tub.	Exterior walls adjacent to showers and tubs shall be insulated.
Electrical/phone box on exterior walls	The air barrier shall be installed behind electrical and communication boxes. Alternatively, air-sealed boxes shall be installed.	—
HVAC register boots	HVAC supply and return register boots that penetrate building thermal envelope shall be sealed to the subfloor, wall covering or ceiling penetrated by the boot.	—
Concealed sprinklers	Where required to be sealed, concealed fire sprinklers shall only be sealed in a manner that is recommended by the manufacturer. Caulking or other adhesive sealants shall not be used to fill voids between fire sprinkler cover plates and walls or ceilings.	—

a. Inspection of log walls shall be in accordance with the provisions of ICC 400.

© 2018 IRC®, International Code Council®

STRAW CONSTRUCTION

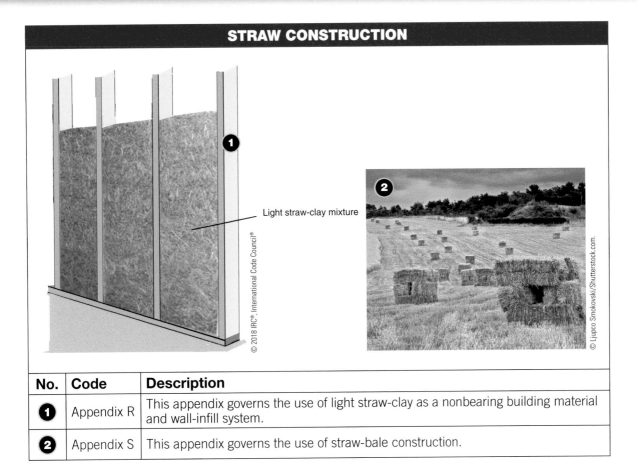

Light straw-clay mixture

© 2018 IRC®, International Code Council®

© Ljupco Smokovski/Shutterstock.com.

No.	Code	Description
1	Appendix R	This appendix governs the use of light straw-clay as a nonbearing building material and wall-infill system.
2	Appendix S	This appendix governs the use of straw-bale construction.

SECTION 2

HVAC

MECHANICAL EQUIPMENT LOCATION

Prefabricated Pad

Concrete Pad

3"

3"

MECHANICAL EQUIPMENT LOCATION *(cont.)*

No.	Code	Description
❶	M1401.2	Heating and cooling equipment and appliances shall be located with respect to building construction and other equipment and appliances to permit maintenance, servicing, and replacement.
❷		Clearances shall be maintained to permit cleaning of heating and cooling surfaces; replacement of filters, blowers, motors, controls, and vent connections; lubrication of moving parts; and adjustments.
❸	M1305.1.3.1	Equipment and appliances supported from the ground shall be level and firmly supported on a concrete slab or other approved material extending not less than 3″ (76 mm) above the adjoining ground. Such support shall be in accordance with the manufacturer's installation instructions. Appliances suspended from the floor shall have a clearance of not less than 6″ (152 mm) from the ground.

You Should Know

- Supports and foundations for the outdoor unit of a heat pump shall be raised at least 3″ above the ground to permit free drainage of defrost water and shall conform to the manufacturer's installation instructions.

HVAC

APPLIANCE ACCESS AND FACTORY-APPLIED NAMEPLATE

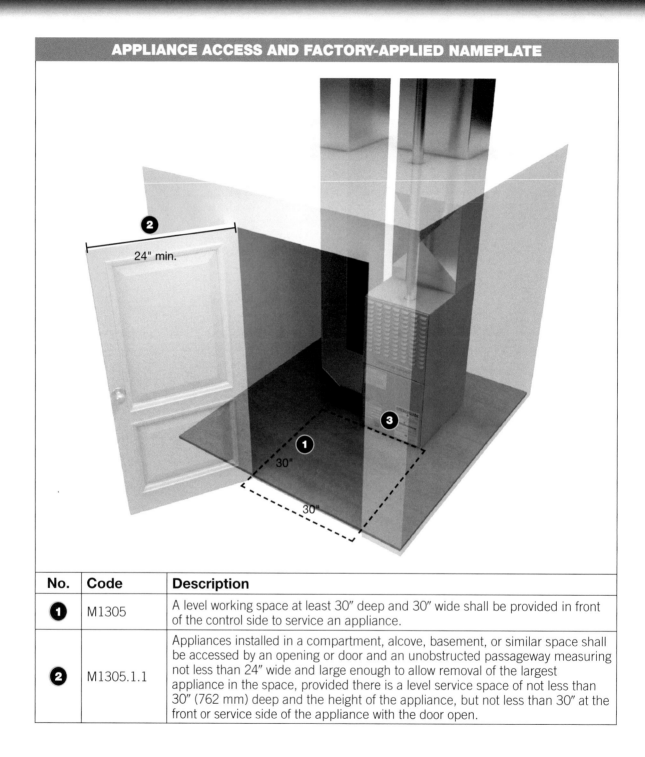

No.	Code	Description
❶	M1305	A level working space at least 30″ deep and 30″ wide shall be provided in front of the control side to service an appliance.
❷	M1305.1.1	Appliances installed in a compartment, alcove, basement, or similar space shall be accessed by an opening or door and an unobstructed passageway measuring not less than 24″ wide and large enough to allow removal of the largest appliance in the space, provided there is a level service space of not less than 30″ (762 mm) deep and the height of the appliance, but not less than 30″ at the front or service side of the appliance with the door open.

APPLIANCE ACCESS AND FACTORY-APPLIED NAMEPLATE *(cont.)*

No.	Code	Description
❸	M1303	A permanent factory-applied nameplate(s) shall be affixed to appliances on which shall appear, in legible lettering, the manufacturer's name or trademark, the model number, a serial number, and the seal or mark of the testing agency. A label shall also include the following: 1. Electrical Appliances: Electrical rating in volts, amperes, and motor phase; identification of individual electrical components in volts, amperes or watts, and motor phase; and in Btu/h (W) output and required clearances 2. Absorption Units: Hourly rating in Btu/h (W), minimum hourly rating for units having step or automatic modulating controls, type of fuel, type of refrigerant, cooling capacity in Btu/h (W), and required clearances 3. Fuel-burning Units: Hourly rating in Btu/h (W), type of fuel approved for use with the appliance, and required clearances 4. Electric Comfort-heating Appliances: The electric rating in volts, amperes, and phase; Btu/h (W) output rating; individual marking for each electrical component in amperes or watts, volts, and phase; required clearances from combustibles 5. Maintenance Instructions: Required regular maintenance actions and title or publication number for the operation and maintenance manual for that particular model and type of product

You Should Know

- Labeling denotes that equipment has had independent testing and evaluation service review and approval.

HVAC

APPLIANCE ACCESS IN ATTICS

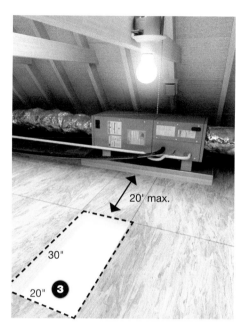

No.	Code	Description
❶	M1305.1.2	Attics containing appliances shall be provided with an opening and a clear and unobstructed passageway large enough to allow removal of the largest appliance, but not less than 30″ high and 22″ wide and not more than 20′ long measured along the centerline of the passageway from the opening to the appliance.
❷		The passageway shall have continuous solid flooring in accordance with Chapter 5 in the IRC code book not less than 24″ wide. A level service space at least 30″ deep and 30″ wide shall be present along all sides of the appliance where access is required.
❸		The clear access opening dimensions shall be a minimum of 20″ by 30″ and large enough to allow removal of the largest appliance.
	M1305.1.3.2	Pit locations: appliances in pits or excavations shall not come into direct contact with the surrounding soil and shall be installed at 3″ above the pit floor.

UNDER-FLOOR ACCESS TO APPLIANCES

No.	Code	Description
❶	M1305.1.3	Under-floor spaces containing appliances shall be provided with an unobstructed passageway large enough to remove the largest appliance, but not less than 30″ (762 mm) high and 22″ (559 mm) wide, nor more than 20′ (6096 mm) long measured along the centerline of the passageway from the opening to the appliance.
❷	M1305.1.3.1	Equipment and appliances supported from the ground shall be level and firmly supported on a concrete slab or other approved material extending not less than 3″ above the adjoining ground. Such support shall be in accordance with the manufacturer's installation instructions. Appliances suspended from the floor shall have a clearance of not less than 6″ from the ground.

You Should Know

- A luminaire controlled by a switch located at the required passageway opening and a receptacle outlet shall be installed at or near the appliance location for both under-floor spaces and attics.

INSTALLATION

TABLE M1306.2 REDUCTION OF CLEARANCES WITH SPECIFIED FORMS OF PROTECTION[a, c, d, e, f, g, h, i, j, k, l]

TYPE OF PROTECTION APPLIED TO AND COVERING ALL SURFACES OF COMBUSTIBLE MATERIAL WITHIN THE DISTANCE SPECIFIED AS THE REQUIRED CLEARANCE WITH NO PROTECTION (See Figures M1306.1 and M1306.2)	WHERE THE REQUIRED CLEARANCE WITH NO PROTECTION FROM APPLIANCE, VENT CONNECTOR, OR SINGLE WALL METAL PIPE IS:									
	36 inches		18 inches		12 inches		9 inches		6 inches	
	Allowable clearances with specified protection (inches)[b] Use column 1 for clearances above an appliance or horizontal connector. Use column 2 for clearances from an appliance, vertical connector, and single-wall metal pipe.									
	Above column 1	Sides and rear column 2	Above column 1	Sides and rear column 2	Above column 1	Sides and rear column 2	Above column 1	Sides and rear column 2	Above column 1	Sides and rear column 2
3½-inch-thick masonry wall without ventilated air space	—	24	—	12	—	9	—	6	—	5
½-inch insulation board over 1-inch glass fiber or mineral wool batts	24	18	12	9	9	6	6	5	4	3
Galvanized sheet steel having a minimum thickness of 0.0236-inch (No. 24 gage) over 1-inch glass fiber or mineral wool batts reinforced with wire or rear face with a ventilated air space	18	12	9	6	6	4	5	3	3	3
3½-inch-thick masonry wall with ventilated air space	—	12	—	6	—	6	—	6	—	6
Galvanized sheet steel having a minimum thickness of 0.0236-inch (No. 24 gage) with a ventilated air space 1-inch off the combustible assembly	18	12	9	6	6	4	5	3	3	2
½-inch-thick insulation board with ventilated air space	18	12	9	6	6	4	5	3	3	3
Galvanized sheet steel having a minimum thickness of 0.0236-inch (No. 24 gage) with ventilated air space over 24 gage sheet steel with a ventilated space	18	12	9	6	6	4	5	3	3	3
1-inch glass fiber or mineral wool batts sandwiched between two sheets of galvanized sheet steel having a minimum thickness of 0.0236-inch (No. 24 gage) with a ventilated air space	18	12	9	6	6	4	5	3	3	3

For SI: 1 inch = 25.4 mm, 1 pound per cubic foot = 16.019 kg/m³, °C = [(°F)-32/1.8], 1 Btu/(h × ft² × °F/in.) = 0.001442299 (W/cm² × °C/cm).

a. Reduction of clearances from combustible materials shall not interfere with combustion air, draft hood clearance and relief, and accessibility of servicing.

b. Clearances shall be measured from the surface of the heat producing appliance or equipment to the outer surface of the combustible material or combustible assembly.

c. Spacers and ties shall be of noncombustible material. No spacer or tie shall be used directly opposite appliance or connector.

d. Where all clearance reduction systems use a ventilated air space, adequate provision for air circulation shall be provided as described. (See Figures M1306.1 and M1306.2.)

e. There shall be at least 1 inch between clearance reduction systems and combustible walls and ceilings for reduction systems using ventilated air space.

f. If a wall protector is mounted on a single flat wall away from corners, adequate air circulation shall be permitted to be provided by leaving only the bottom and top edges or only the side and top edges open with at least a 1-inch air gap.

g. Mineral wool and glass fiber batts (blanket or board) shall have a minimum density of 8 pounds per cubic foot and a minimum melting point of 1,500°F.

h. Insulation material used as part of a clearance reduction system shall have a thermal conductivity of 1.0 Btu inch per square foot per hour °F or less. Insulation board shall be formed of noncombustible material.

i. There shall be at least 1 inch between the appliance and the protector. In no case shall the clearance between the appliance and the combustible surface be reduced below that allowed in this table.

j. All clearances and thicknesses are minimum; larger clearances and thicknesses are acceptable.

k. Listed single-wall connectors shall be permitted to be installed in accordance with the terms of their listing and the manufacturer's instructions.

l. For limitations on clearance reduction for solid-fuel-burning appliances see Section M1306.2.1.

© 2018 IRC® International Code Council®

INSTALLATION *(cont.)*	
Code	**Description**
M1306.1	Appliances shall be installed with the clearances from unprotected combustible materials as indicated on the appliance label and in the manufacturer's installation instructions.
M1306.2	Reduction of clearances shall be in accordance with the appliance manufacturer's instructions and Table M1306.2.1. The allowable clearance shall be based on an approved reduced clearance protective assembly that is listed and labeled in accordance with UL 1618. Forms of protection with ventilated air space shall conform to the following requirements: 1. Not less than 1″ air space shall be provided between the protection and combustible wall surface. 2. Air circulation shall be provided by having edges of the wall protection open at least 1″. 3. If the wall protection is mounted on a single flat wall away from corners, air circulation shall be provided by having the bottom and top edges, or the side and top edges, open at least 1″. 4. Wall protection covering two walls in a corner shall be open at the bottom and top edges at least 1″.

HVAC

You Should Know

- Table M1306.2 shall not be used to reduce the clearance required for solid-fuel appliances listed for installation with minimum clearances of 12″ (305 mm) or less.
- For solid-fuel appliances listed for installation with minimum clearances greater than 12″, Table M1306.2 shall not be used to reduce the clearance to less than 12″.

APPLIANCES IN GARAGES

No.	Code	Description
❶	M1307.3	Appliances having an ignition source shall be elevated such that the source of ignition is not less than 18″ above the floor in garages. For the purpose of this section, rooms or spaces that are not part of the living space of a dwelling unit and that communicate with a private garage through openings shall be considered to be part of the garage. Appliances listed as flammable-vapor, ignition-resistant do not require elevation.
❷	M1307.3.1	Appliances shall not be installed in a location subject to vehicle damage except where protected by approved barriers.
❸	M1308.2	Where piping will be concealed within light frame construction assembly, the piping shall be protected against penetration by fasteners in accordance with Sections M1308.2.1 through M1308.2.3.
❹	M1307.1	Installation of appliances shall conform to the conditions of their listing and label and the manufacturer's installation instructions. The manufacturer's operating and installation instructions shall remain attached to the appliance.
❺	M1307.2	Appliances designed to be fixed in position shall be fastened or anchored in an approved manner. In Seismic Design Categories D^1 and D^2, water heaters and thermal storage units shall be anchored or strapped to resist horizontal displacement caused by earthquake motion.

You Should Know

- Required protective steel shield plates in item 3 above must be at least 0.0575″ thick (16 gauge).

HEATING AND COOLING EQUIPMENT

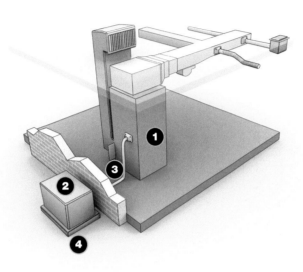

No.	Code	Description
1	M1401.3	Heating and cooling equipment and appliances shall be sized in accordance with *ACCA Manual S* based on building loads calculated in accordance with *ACCA Manual J* or other approved heating and cooling calculation methodologies.
2	M1401.4	Equipment installed outdoors shall be listed and labeled for outdoor installation. Supports and foundations shall prevent excessive vibration, settlement, or movement of the equipment. Supports and foundations shall be in accordance with Section M1305.1.3.1.
3	M1402	Clearances for central furnaces shall be provided in accordance with the listing and the manufacturer's installation instructions. Combustion air shall be supplied in accordance with Chapter 17 in the IRC code book. Combustion air openings shall be unobstructed for a distance of not less than 6″ in front of the openings.
4	M1403.1	Electric heat pumps shall be listed and labeled in accordance with UL 1995 or UL/CSA/ANCE 60335-2-40.

MANUAL J HEAT CALCULATION

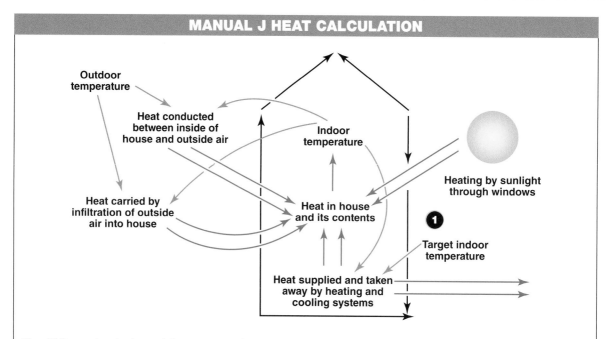

The IRC mechanical provisions set out the standard for calculating the need for heating and cooling a home. The method prescribed by the IRC is in a standard developed by the Air Conditioning Contractors of America (ACCA). The standard is *ACCA Manual J.*

Manual J is a guideline for calculating the heat losses and gains in a building. In order to do this effectively, several factors must be considered, including the geographic zone for the building, the design temperature of the building, and the way the building's components are constructed. This last factor is important because it identifies the heat flow through the building component (wall, floor, or roof).

When the conditions for heat transfer are known, the proper HVAC systems can be determined. It is important to select the right size. The Goldilocks approach is important because too small equipment will not provide adequate heat (or cooling) and too large equipment will result in inefficiencies and overconsumption. The proper-size cooling and heating system will optimize temperature control, humidity, ventilation, and air movement. Using the *Manual J* method, the determination of heat flow (in or out of a building) follows methodology established by engineering principles. Here's the basic equation for heat flow:

$$Q = \mu A / \Delta T$$

Q = heat flow

μ = the μ value of a particular material

ΔT = the difference between the outside temperature and the inside design temperature

Essentially, you break down a wall section into components and find the overall heat flow for that wall section. The ACCA standard does that work for you. When you obtain the total loss of heat, you can simply match a heating (or cooling) system for the building.

MANUAL J HEAT CALCULATION *(cont.)*

Table 1: Weather Data: This table provides weather and temperature conditions based on state and city.

Table 2: Heat Loss: This table displays heat transfer multipliers (HTM) necessary for calculating the heating load based on construction materials.

Table 3: Heat Gain Through Glass: This table displays HTM necessary for calculating the cooling load based on construction materials, number of glazing panes, inside and outside shading, and the orientation of the window (which way it is facing).

Table 4: Heat Gain: This table displays HTM necessary for calculating the cooling load based on construction materials other than glazing.

No.	Code	Description
❶	M1401.3	Heating and cooling equipment shall be sized in accordance with *ACCA Manual S* or other approved sizing methodologies based on building loads calculated in accordance with *ACCA Manual J* or other approved heating and cooling calculation methodologies.

HVAC

SAMPLE HEAT LOSS CALCULATION

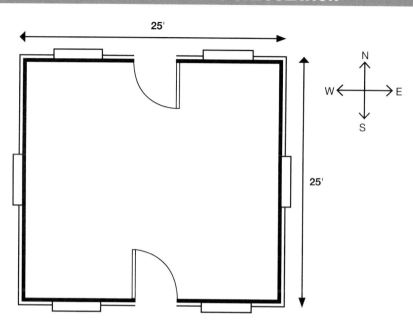

- Location: Norfolk, Virginia
- All windows are 3′ × 5′, clear glass, double pane with a metal frame.
  - There is no internal shading.
- Walls are wood frame with R-19 batt insulation and ½″ drywall on interior walls.
- Ceiling is 8′ high.
- Attic has R-30 batt insulation (dark colored).
- Doors are $3^0 \times 6^8$ wood with solid core and a metal storm door.
- Indoor design temperature for winter is 68°F (minimum code requirement).
- Outdoor design temperature for summer is 75°F.

The size of heating and cooling equipment is based on the heat loss and heat gain calculations.

Step	Description
1	Search for weather data for Norfolk, Virginia, in Table 1A.

	Latitude	99% Design Db	1% Design Db	Design Grains 55%	Design Grains 50%	Daily Range
1	36	24	91	42	49	M

Step	Description
2	Determine the temperature difference between outside air and inside design temperature. Winter inside temperature (68°F) – Outside 99% Db (20) = 48°F. Summer outside 1% Db (91°F) – Summer inside temperature (75 °F) = 16°F. Winter temperature difference = 48°F. Summer temperature difference = 16°F.
3	Heat loss = Area of wall, floor, or roof × HTM (heat transfer multiplier) in Btu/h.
4	Heat gain = Area of wall, floor, or roof × HTM (heat transfer multiplier) in Btu/h.

SAMPLE HEAT LOSS CALCULATION *(cont.)*

FIRST SAMPLE: HEAT LOSS THROUGH WALLS CALCULATION

North Wall
25' length × 8' wall height = 200 ft²
200 ft² − 15 ft² − 20 ft² = 165 ft²
165 ft² × 3.0 (Heat Transfer Multiplier) = 495 Btu/h

East Wall
25' length × 8' wall height = 200 ft²
200 ft² − 15 ft² = 185 ft²
185 ft² × 3.0 (Heat Transfer Multiplier) = 555 Btu/h

West Wall
25' length × 8' wall height = 200 ft²
200 ft² − 15 ft² = 185 ft²
185 ft² × 3.0 (Heat Transfer Multiplier) = 555 Btu/h

South Wall
25' length × 8' wall height = 200 ft²
200 ft² − 15 ft² − 15 ft² − 20 ft² = 150 ft²
150 ft² × 3.0 (Heat Transfer Multiplier) = 450 Btu/h

North Windows
2 windows × 3' × 5' = 30 ft²
30 ft² × 36.3 = 1089 Btu/h

East Windows
1 window × 3' × 5' = 15 ft²
15 ft² × 36.3 Btu/h = 544 Btu/h

West Windows
1 window × 3' × 5' = 15 ft²
15 ft² × 36.3 Btu/h = 544 Btu/h

South Windows
2 windows × 3' × 5' = 30 ft²
30 ft² × 36.3 = 1089 Btu/h

North Door
20 ft² × 16 Btu/h − 320 Btu/h

South Door
20 ft² × 16 Btu/h − 320 Btu/h

Total:
495 + 555 + 555 + 450 + 1089 + 544 + 544 + 1089 + 320 + 320 = 5961 Btu/h

This means that for the four walls, almost 6000 Btus are lost every hour. Of course, the roof and floor must be considered for a complete evaluation of the building. But when that is complete, the total Btu/h demand is used to select equipment size. A proper selection would be slightly higher.

SAMPLE HEAT LOSS CALCULATION (cont.)

SECOND SAMPLE: HEAT LOSS CALCULATION

A. Heat loss through walls facing each direction (*excluding doors and windows*)

North Wall:
25' *(length)* × 8' *(ceiling height)* = 200 sq. ft.
200 sq. ft. − 30 sq. ft. *(window take-off)* = 170 sq. ft.
170 sq. ft. − 20 sq. ft. *(door take-off)* = 150 sq. ft.
150 sq. ft. × 3.0 HTM *(Table 2, #12 H)* = 450 Btu/hr

South Wall:
25' *(length)* × 8' *(ceiling height)* = 200 sq. ft.
200 sq. ft. − 30 sq. ft. *(window take-off)* = 170 sq. ft.
170 sq. ft. − 20 sq. ft. *(door take-off)* = 150 sq. ft.
150 sq. ft. × 3.0 HTM *(Table 2, #12 H)* = 450 Btu/hr

East Wall:
25' *(length)* × 8' *(ceiling height)* = 200 sq. ft.
200 sq. ft. − 15 sq. ft. *(window take-off)* = 185 sq. ft.
185 sq. ft. × 3.0 HTM *(Table 2, #12 H)* = 555 Btu/hr

West Wall:
25' *(length)* × 8' *(ceiling height)* = 200 sq. ft.
200 sq. ft. − 15 sq. ft. *(window take-off)* = 185 sq. ft.
185 sq. ft. × 3.0 HTM *(Table 2, #12 H)* = 555 Btu/hr

No. 12 Wood Frame Exterior Walls with Sheathing and Siding or Brick, or Other Exterior Finish.	Winter Temperature Difference												
	20	25	30	35	40	45	50	55	60	65	70	75	80
Cavity Insul Sheathing	HTM (Btu/h per sq. ft.)												
A. None ½" Gypsum Brd (R-0.5)	5.4	6.8	8.1	9.5	10.8	12.2	13.6	14.9	16.3	17.6	19.0	20.3	
B. None ½" Asphalt Brd (R-1.3)	4.3	5.4	6.5	7.6	8.7	9.8	10.8	11.9	13.0	14.1	15.2	16.3	
C. R-11 ½" Gypsum (R-0.5)	1.8	2.3	2.7	3.1	3.6	4.0	4.5	4.9	5.4	5.8	6.3	6.7	
D. R-11 ½" Asphalt Brd (R-1.3) R-11 ½" Bead Brd (R-1.8) R-13 ½" Gypsum Brd (R-0.5)	1.6	2.0	2.4	2.8	3.2	3.6	4.0	4.4	4.8	5.2	5.6		
E. R-11 ½" Extr Poly Brd (R-2.5) R-11 ¾" Bead Brd (R-2.7) R-13 ½" Asphalt Brd (R-1.3) R-13 ¾" Bead Brd (R-1.8)	1.5	1.9	2.3	2.6	3.0	3.4	3.8	4.1	4.5	4.9	5.3		
F. R-11 1" Bead Brd (R-3.6) R-11 ¾" Extr Poly Brd (R-3.8) R-13 ½" Extr Poly Brd (R-2.5) R-13 ¾" Bead Brd (R-2.7)	1.4	1.8	2.1	2.4	2.8	3.2	3.5	3.8	4.2	4.6			
G. R-13 ¾" Extr Poly Brd (R-3.8) R-13 1" Bead Brd (R-3.6)	1.3	1.6	2.0	2.3	2.6	2.9	3.3	3.6	3.9	4.2			
H. R-11 1" Extr Poly Brd (R-5.0) R-13 1" Extr Poly Brd (R-5.0) R-19 ½" Gypsum Brd (R-0.5)	1.2	1.5	1.8	2.1	2.4	2.7	3.0	3.3	3.6				
I. R-19 ½" Asphalt Brd (R-1.3) R-19 ½" Bead Brd (R-1.8)	1.1	1.4	1.6	1.9	2.2	2.5	2.8	3.0	3.3				
J. R-11 R-8 Sheathing R-13 R-8 Sheathing R-19 ½" or ¾" Extr Poly Brd R-19 ¾" or 1" Bead Brd	1.0	1.3	1.5	1.7	2.0	2.2	2.5	2.7					
K. R-19 1" Extr Poly Brd (R-5.0)	.9	1.1	1.3	1.6	1.8	2.0	2.2	2.5					
L. R-19 R-8 Sheathing	.8	1.0	1.2	1.4	1.6	1.8	2.0						
M. R-27 Wall	.7	.9	1.1	1.3	1.5	1.7	1.9						
N. R-30 Wall	.7	.8	1.0	1.2	1.3	1.5	1.5						
O. R-33 Wall	.6	.8	.9	1.1	1.2	1.4	1.5						

Total Heat Loss Through Walls = 2010 Btu/hr

SAMPLE HEAT LOSS CALCULATION *(cont.)*

B. Heat loss through windows facing each direction

North Windows: 2 *(windows)* × 3′ × 5′ = 30 sq. ft.
30 sq. ft. × 36.3 HTM *(Table 2, #3 C)* = 1089 Btu/hr

South Windows: 2 *(windows)* × 3′ × 5′ = 30 sq. ft.
30 sq. ft. × 36.3 HTM *(Table 2, #3 C)* = 1089 Btu/hr

East Window: 1 *(window)* × 3′ × 5′ = 15 sq. ft.
15 sq. ft. × 36.3 HTM *(Table 2, #3 C)* = 544.5 Btu/hr

West Window: 1 *(window)* × 3′ × 5′ = 15 sq. ft.
15 sq. ft. × 36.3 HTM *(Table 2, #3 C)* = 544.5 Btu/hr

No. 3 Double Pane Window	Winter Temperature Difference											
	20	25	30	35	40	45	50	55	60	65	70	75
	HTM (Btu/h per sq. ft.)											
Clear Glass	21.7											
A. Wood Frame	11.0	13.8	16.5	19.3	22.0	24.8	27.6	30.3	33.1	35.8	38.6	41.3
B. T.I.M. Frame	12.2	15.2	18.3	21.3	24.4	27.4	30.5	33.5	36.5	39.6	42.6	
C. Metal Frame	14.5	18.1	21.8	25.4	29.0	32.6	36.3	39.9	43.5	47.1	50.8	
Low Emittance Glass												
D. Wood Frame	7.2	9.0	10.8	12.6	14.4	16.2	18.1	19.9	21.7	23.5	25.3	
E. T.I.M. Frame	8.0	10.0	12.0	14.0	16.0	18.0	20.0	21.9	23.9	25.9	27	
F. Metal Frame	9.5	11.9	14.3	16.6	19.0	21.4	23.8	26.1	28.5	30		
Adjustable Blind Between Panes												
G. Wood Frame	4.8	5.9	7.1	8.3	9.5	10.7	11.9	13.1	14.3	15.4		
H. T.I.M. Frame	5.3	6.6	7.9	9.2	10.5	11.8	13.1	14.4	15.8	17.1		
Total Heat Loss Through Windows = 3267 Btu/hr												

C. Heat loss through doors facing each direction *(convert inches into feet)*

North Door: 3′ × 6.67′ (8″ ÷ 12″ = .67′) = 20.01 sq. ft. or 20 sq. ft.
20 sq. ft. × 16 HTM *(Table 2, #10 F)* = 320 Btu/hr

South Door: 3′ × 6.67′ (8″ ÷ 12″ = .67′) = 20.01 sq. ft. or 20 sq. ft.
20 sq. ft. × 16 HTM *(Table 2, #10 F)* = 320 Btu/hr

No. 10 Wood Doors	Winter Temperature Difference											
	20	25	30	35	40	45	50	55	60	65	70	75
	HTM (Btu/h per sq. ft.)											
A. Hollow Core	11.2	14.0	16.8	19.6	22.4	25.2	28.0	30.8	33.6	36.4	39.2	42.0	44.
B. Hollow Core and Wood Storm	6.6	8.3	9.9	11.6	13.2	14.9	16.5	18.2	19.8	21.5	23.1	24.8	26.4
C. Hollow Core and Metal Storm	7.2	9.0	10.8	12.6	14.4	16.2	18.0	19.8	21.6	23.4	25.2	27.0
D. Solid Core	9.2	11.5	13.8	16.1	18.4	20.7	23.0	25.3	27.6	29.9	32.2	
E. Solid Core and Wood Storm	5.8	7.3	8.7	10.2	11.6	13.1	14.5	16.0	17.4	18.9	20.3	
F. Solid Core and Metal Strom	6.4	8.0	9.6	11.2	12.8	14.4	16.0	17.6	19.2	20.8	22.4	
G. Panel	13.4	16.8	20.1	23.5	26.8	30.2	33.5	36.9	40.2	43.6	46	
H. Panel and Wood Storm	7.2	9.0	10.8	12.6	14.4	16.2	18.0	19.8	21.6	23.4		
I. Panel and Metal Storm	8.2	10.3	12.3	14.4	16.4	18.5	20.5	22.6	24.6	26.7		
Total Heat Loss Through Doors = 640 Btu/hr												

HVAC

SAMPLE HEAT LOSS CALCULATION *(cont.)*

D. Heat loss through ceiling

Condensate and Refrigerant

Ceiling: 25' × 25' = 625 sq. ft.
625 sq. ft. × 1.6 HTM *(Table 2, #16 G)* = 1000 Btu/hr

No. 16 Ceiling Under a Ventilated Attic Space or Unheated Room	Winter Temperature Difference											
	20	25	30	35	40	45	50	55	60	65	70	75
	HTM (Btu/h per sq. ft.)											
A. No Insulation	12.0	15.0	18.0	21.0	24.0	27.0	29.9	32.9	35.9	38.9	41.9	44.9
B. R-7 Insulation	2.4	3.0	3.6	4.2	4.8	5.4	6.0	6.6	7.2	7.8	8.4	9.0
C. R-11 Insulation	1.8	2.2	2.6	3.1	3.5	4.0	4.4	4.8	5.3	5.7	6.2	6.6
D. R-19 Insulation	1.1	1.3	1.6	1.9	2.1	2.4	2.6	2.9	3.2	3.4	3.7	
E. R-22 Insulation	1.0	1.2	1.4	1.7	1.9	2.2	2.4	2.6	2.9	3.1	3.4	
F. R-26 Insulation	.8	1.0	1.1	1.3	1.5	1.7	1.9	2.1	2.3	2.5	2.7	
G. R-30 Insulation	.7	.8	1.0	1.2	1.3	1.5	1.6	1.8	2.0	2.1	2.3	
H. R-38 Insulation	.5	.7	.8	.9	1.0	1.2	1.3	1.4	1.6	1.7		
I. R-44 Insulation	.5	.6	.7	.8	.9	1.0	1.1	13	1.4	1.5		
J. R-57 Insulation	.3	.4	.5	.6	.7	.8	.8	.9	1.0	1.1		
K. Wood Decking. No Insulation	5.7	7.2	8.6	10.0	11.5	12.9	14.3	15.7	17.2	18.6		

Total Heat Loss Through Ceiling = 1000 Btu/hr

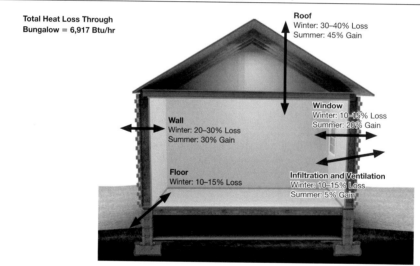

Total Heat Loss Through Bungalow = 6,917 Btu/hr

Roof
Winter: 30–40% Loss
Summer: 45% Gain

Window
Winter: 10–15% Loss
Summer: 20% Gain

Wall
Winter: 20–30% Loss
Summer: 30% Gain

Floor
Winter: 10–15% Loss

Infiltration and Ventilation
Winter: 10–15% Loss
Summer: 5% Gain

Code	Description
M1411.3	Condensate disposal: Condensate from all cooling coils or evaporators shall be conveyed from the drain pan outlet to an approved place of disposal. Such piping shall maintain a minimum horizontal slope in the direction of discharge of not less than ⅛ unit vertical in 12 units horizontal. Condensate shall not discharge into a street, alley, or other areas where it would cause a nuisance.
M1411.8	Locking access port caps: Refrigerant circuit access ports located outdoors shall be fitted with locking-type, tamper-resistant caps or shall be otherwise secured to prevent unauthorized access.

CLOTHES DRYER EXHAUST

TABLE M1502.4.5.1 DRYER EXHAUST DUCT FITTING EQUIVALENT LENGTH

DRYER EXHAUST DUCT FITTING TYPE	EQUIVALENT LENGTH
4 inch radius mitered 45 degree elbow	2 feet 6 inches
4 inch radius mitered 90 degree elbow	5 feet
6 inch radius smooth 45 degree elbow	1 foot
6 inch radius smooth 90 degree elbow	1 foot 9 inches
8 inch radius smooth 45 degree elbow	1 foot
8 inch radius smooth 90 degree elbow	1 foot 7 inches
10 inch radius smooth 45 degree elbow	9 inches
10 inch radius smooth 90 degree elbow	1 foot 6 inches

For SI: 1 inch = 25.4 mm, 1 foot = 304.8 mm, 1 degree = 0.0175 rad.

HVAC

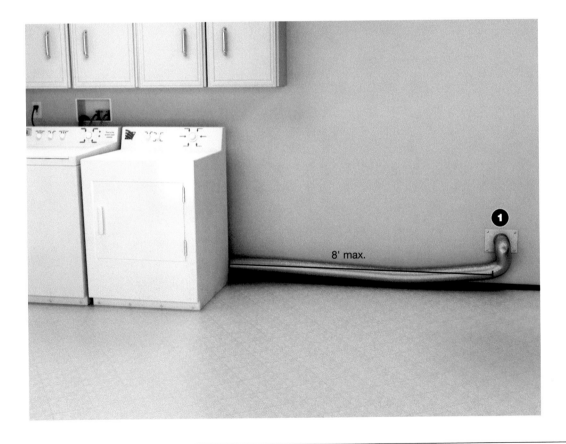

8' max.

CLOTHES DRYER EXHAUST *(cont.)*

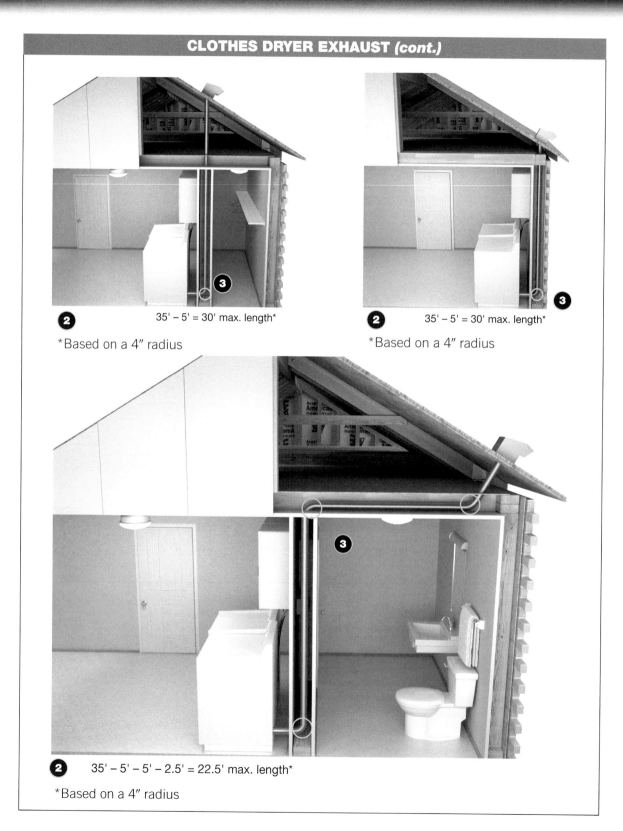

2 35' – 5' = 30' max. length*

*Based on a 4″ radius   **3**

2 35' – 5' = 30' max. length*

*Based on a 4″ radius   **3**

3

2 35' – 5' – 5' – 2.5' = 22.5' max. length*

*Based on a 4″ radius

CLOTHES DRYER EXHAUST *(cont.)*

No.	Code	Description
❶	M1502.3	Exhaust ducts shall terminate on the outside of the building. Exhaust duct terminations shall be in accordance with the dryer manufacturer's installation instructions. If the manufacturer's instructions do not specify a termination location, the exhaust duct shall terminate not less than 3' (914 mm) in any direction from openings into buildings. Exhaust duct terminations shall be equipped with a back-draft damper. Screens shall not be installed at the duct termination.
	M1502.3.1	The passageway of dryer exhaust duct terminals shall not be diminished in size and shall provide an open area of not less than 12.5 square inches.
❷	M1502.4.1	Exhaust ducts shall have a smooth interior finish and shall be constructed of metal a minimum 0.0157" thick or #28 gauge. The exhaust duct size shall be 4" nominal in diameter.
❸	M1502.4.5.1	The maximum length of the exhaust duct shall be 35' from the connection to the transition duct from the dryer to the outlet terminal. Where fittings are used, the maximum length of the exhaust duct shall be reduced in accordance with Table M1502.4.4.1. The size and maximum length of exhaust duct could also be according to dryer manufacturer's installation instructions. See Section M1502.4.5.2.
	M1502.4.2	Duct installation: Exhaust ducts shall be supported at intervals not to exceed 12' and shall be secured in place. The insert end of the duct shall extend into the adjoining duct or fitting in the direction of airflow. Exhaust duct joints shall be sealed in accordance with Section M1601.4.1 in the IRC book and shall be mechanically fastened. Ducts shall not be joined with screws or similar fasteners that protrude more than ⅛" into the inside of the duct.
	M1502.4.6	Where the exhaust duct exceeds 35' in equivalent length, the length shall be identified on a permanent label or tag. The label or tag shall be located within 6' of the exhaust duct connection.

HVAC

You Should Know

- Where space for a clothes dryer is provided, an exhaust duct system shall be installed.
- Where the clothes dryer is not installed at the time of occupancy, the exhaust duct shall be capped or plugged in the space in which it originates and identified and marked "future use."

RANGE HOODS AND BATHROOM FANS

RANGE HOODS AND BATHROOM FANS *(cont.)*

No.	Code	Description
❶	M1503.2	Where domestic cooking exhaust equipment is provided, it shall comply with one of the following: 1. Fan for overhead range hoods and downdraft exhaust equipment not integral with cooking appliances shall be listed with UL507. 2. Overhead range hoods and downdraft exhaust equipment with integral fans shall be listed with UL507. 3. Domestic cooking appliances with integral downdraft exhaust equipment shall be listed and labeled with ANSI Z21.1 or UL 858. 4. Microwave ovens with integral exhaust for installation over the cooking surface shall be listed and labeled with UL923.
❷	M1503.2.1	Domestic open top broiler units shall be provided with a metal exhaust hood having a thickness of not less than 0.0157″ (no. 28 gauge). Such hood must be installed according to Section M1503.2.1.
	M1505.2	Exhaust air from bathrooms and toilet rooms shall not be recirculated within a residence or to another dwelling unit and shall be exhausted directly to the outdoors. Exhaust air from bathrooms and toilet rooms shall not discharge into an attic, crawl space, or other areas inside the building.
❸	M1503.1	The installation of a listed and labeled cooking appliance or microwave oven over a listed and labeled cooking appliance shall conform to the terms of the upper appliance's listing and label and the manufacturer's installation instructions and ANSI Z21.1 or UL858.
❹	M1503.2.1	Domestic open-top broiler units shall have a metal exhaust hood, having a minimum thickness of 0.0157″ (No. 28 gauge) with ¼″ clearance between the hood and the underside of combustible material or cabinets. A clearance of at least 24″ shall be maintained between the cooking surface and the combustible material or cabinet.

You Should Know

- Where one or more fuel-burning appliances are neither direct-vent nor mechanical draft vented, makeup air is required for cooking exhaust exceeding 400 cfm per Section M1503.6.

- Mechanical air intake openings must be 10′ from exhaust air openings unless the exhaust is 3′ above the air intake per Section M1504.3, item 3.

EXHAUST SYSTEMS

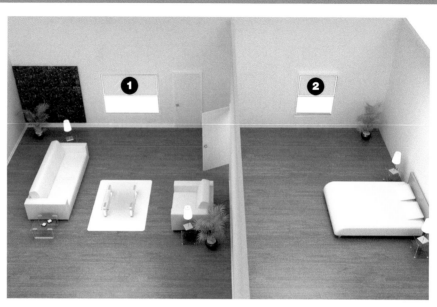

No.	Code	Description
1	R303.1	All habitable rooms shall have an aggregate glazing area of not less than 8% of the floor area of such rooms. Natural ventilation shall be through windows, doors, louvers, or other approved openings to the outdoor air. Such openings shall be provided with ready access or shall otherwise be readily controllable by the building occupants. Glazed areas are not required where artificial lighting and mechanical ventilation are provided.
2	M1505.4.4	Local exhaust rates shall be designed to have the capacity to exhaust the minimum airflow rate determined in accordance with Table M1505.4.4.
	M1504.3	Exhaust openings: Air exhaust openings shall terminate not less than 3' from property lines, 3' from operable or nonoperable openings into the building, and 10' from mechanical air intakes except where the opening is located 3' above the air intake. Openings shall comply with Sections R303.5.2 and R303.6 in the IRC code book.
	M1505.4 and R303.4	Whole-house mechanical ventilation: The whole-house ventilation system shall consist of one or more supply or exhaust fans or a combination of such and associated ducts and controls.

You Should Know

- Section R303 in the IRC code book requires a whole-house mechanical ventilation system when the air changes due to infiltration is less than 5 air changes per hour (5 ACH). This section M1505.4 defines how this system is to be installed.

ABOVEGROUND DUCT SYSTEMS

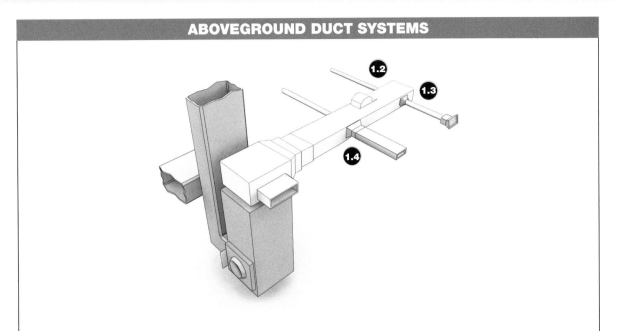

No.	Code	Description
❶	M1601.1.1	Aboveground duct systems shall conform to the following: 1. Equipment connected to duct systems shall be designed to limit discharge air temperature to a maximum of 250°F. 2. Factory-made air ducts shall be listed and labeled in accordance with UL 181 and manufacturer's instructions. 3. Fibrous duct construction shall conform to the SMACNA *Fibrous Glass Duct Construction Standards* or NAIMA *Fibrous Glass Duct Construction Standards*. 4. Field fabricated and shop fabricated metal and flexible duct construction shall conform to the SMACNA HVAC *Fibrous Glass Duct Construction* standards or *NAIMA Fibrous Glass Duct Construction* standards. 5. Use of gypsum products to construct return air ducts or plenums is permitted, provided that the air temperature does not exceed 125°F (52°C) and exposed surfaces are not subject to condensation. 6. Duct systems shall be constructed of materials having a flame spread index not greater than 200. 7. Stud wall cavities and the spaces between solid floor joists to be used as air plenums shall comply with the following conditions: 1. These cavities or spaces shall not be used as a plenum for supply air. 2. These cavities or spaces shall not be part of a required fire-resistance-rated assembly. 3. Stud wall cavities shall not convey air from more than one floor level. 4. Stud wall cavities and joist-space plenums shall be isolated from adjacent concealed spaces by tight-fitting fire-blocking in accordance with Section R602.8 in the IRC code book. 5. Stud wall cavities in the outside walls of the building envelope assemblies shall not be utilized as air plenums.

HVAC

UNDERGROUND DUCT SYSTEMS

Code	Description
M1601.1.2	Underground duct systems must be made of approved concrete, clay, metal, or plastic. The maximum design temperature for plastic ducts is 150°F.
M1601.2	Vibration isolators installed between equipment and ducts must not exceed 10″.
M1601.4.1	All longitudinal and transverse joints, seams, and connections in metallic and nonmetallic ducts shall be construction as specified in SMACNA *HVAC Duct Construction Standards* and NAIMA *Fibrous Glass Duct Construction Standards*. All joints shall be securely fastened and sealed with welds, gaskets, mastics (adhesives), mastic-plus-embedded-fabric systems, or tapes.
	Crimp joints for round metallic ducts shall have a contact lap of not less than 1″ and shall be mechanically fastened by means of at least three sheet-metal screws or rivets equally spaced around the joint.
M1601.4.4	Ducts shall be supported in accordance with the manufacturer's installation instructions or the referenced SMACNA standard as applicable.

You Should Know

- Closure systems used to seal flexible air ducts and flexible air connectors shall comply with UL 181B and shall be marked "181 B-FX" for pressure sensitive tape or "181 BM" for mastic.

DUCT INSTALLATION

TABLE M1601.1.1 DUCT CONSTRUCTION MINIMUM SHEET METAL THICKNESS FOR SINGLE DWELLING UNITS[a]

ROUND DUCT DIAMETER (inches)	STATIC PRESSURE			
	½ inch water gage		1 inch water gage	
	Thickness (inches)		Thickness (inches)	
	Galvanized	Aluminum	Galvanized	Aluminum
≤ 12	0.013	0.018	0.013	0.018
12 to 14	0.013	0.018	0.016	0.023
15 to 17	0.016	0.023	0.019	0.027
18	0.016	0.023	0.024	0.034
19 to 20	0.019	0.027	0.024	0.034

RECTANGULAR DUCT DIMENSION (inches)	STATIC PRESSURE			
	½ inch water gage		1 inch water gage	
	Thickness (inches)		Thickness (inches)	
	Galvanized	Aluminum	Galvanized	Aluminum
≤ 8	0.013	0.018	0.013	0.018
9 to 10	0.013	0.018	0.016	0.023
11 to 12	0.016	0.023	0.019	0.027
13 to 16	0.019	0.027	0.019	0.027
17 to 18	0.019	0.027	0.024	0.034
19 to 20	0.024	0.034	0.024	0.034

For SI: 1 inch = 25.4 mm, 1 inch water gage = 249 Pa.

a. Ductwork that exceeds 20 inches by dimension or exceeds a pressure of 1 inch water gage (250 Pa) shall be constructed in accordance with SMACNA *HVAC Duct Construction Standards—Metal and Flexible.*

© 2018 IRC®, International Code Council®

HVAC

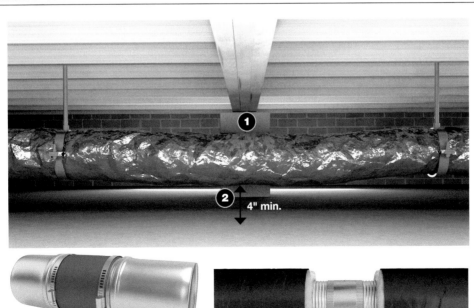

CT INSTALLATION (cont.)

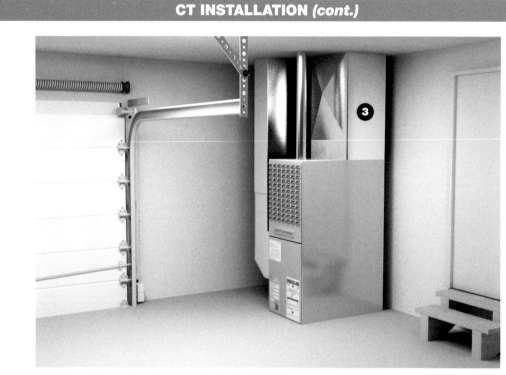

No.	Code	Description
	M1601.1.1, Item #4	Minimum thickness of metal duct material shall be as listed in Table M1601.1.1. Galvanized steel shall conform to ASTM A 653. Metallic ducts shall be fabricated in accordance with SMACNA *Duct Construction Standards Metal and Flexible.*
❶	M1601.4.7	Factory-made air ducts shall not be installed in or on the ground, in tile or metal pipe, or within masonry or concrete.
❷	M1601.4.8	Ducts shall be installed with at least a 4″ (102 mm) separation from earth except underground systems.
❸	M1601.4.9 and R302.5.2	Ducts in the garage and ducts penetrating the walls or ceilings separating the dwelling from the garage shall be constructed of a minimum No. 26 gauge sheet steel or other approved material and shall have no openings into the garage.

You Should Know

- Duct installation involves proper closure systems. These are regulated by a common UL standard UL181. However, there are many subcategories for various types of duct material. Each of these types must be marked with the proper standard for the specific duct material.

- Ducts in flood hazard areas must be installed to prevent water from entering and accumulating and to resist hydrodynamic loads and stresses and installed per Section R322.1.6.

RETURN AIR

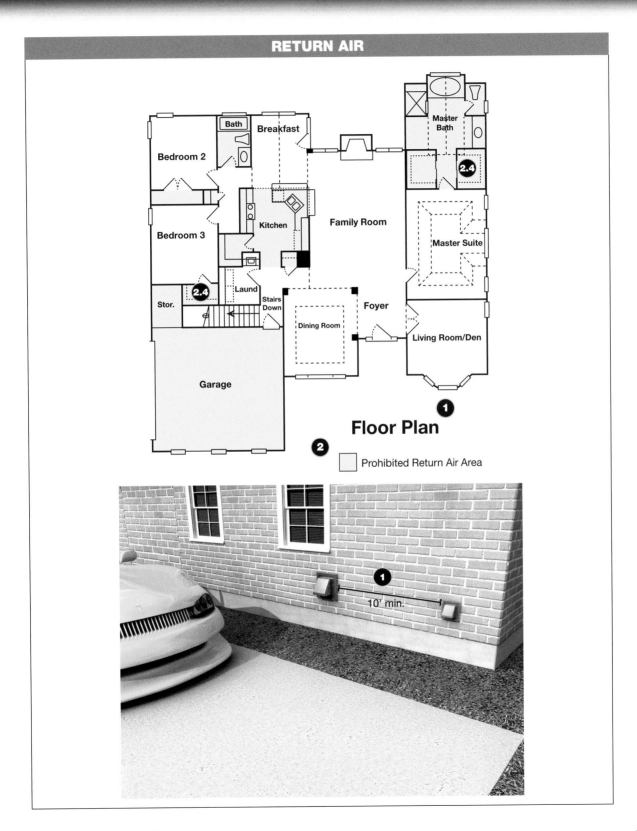

Floor Plan

Prohibited Return Air Area

10' min.

HVAC

RETURN AIR *(cont.)*

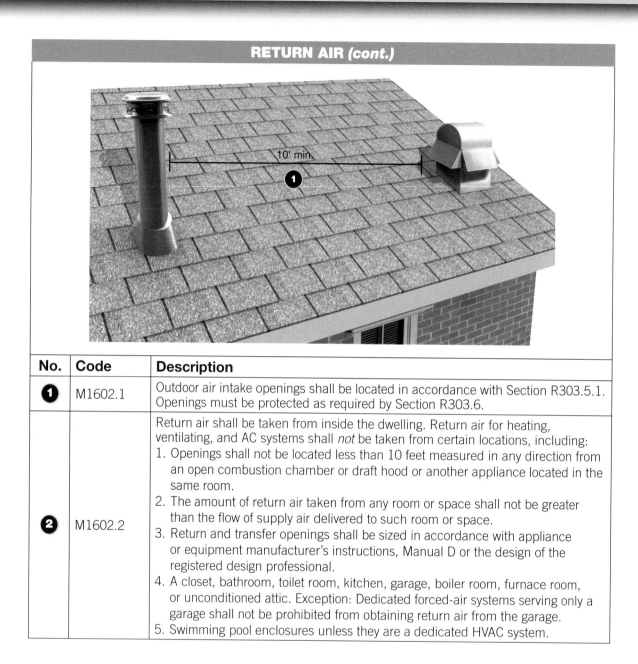

No.	Code	Description
❶	M1602.1	Outdoor air intake openings shall be located in accordance with Section R303.5.1. Openings must be protected as required by Section R303.6.
❷	M1602.2	Return air shall be taken from inside the dwelling. Return air for heating, ventilating, and AC systems shall *not* be taken from certain locations, including: 1. Openings shall not be located less than 10 feet measured in any direction from an open combustion chamber or draft hood or another appliance located in the same room. 2. The amount of return air taken from any room or space shall not be greater than the flow of supply air delivered to such room or space. 3. Return and transfer openings shall be sized in accordance with appliance or equipment manufacturer's instructions, Manual D or the design of the registered design professional. 4. A closet, bathroom, toilet room, kitchen, garage, boiler room, furnace room, or unconditioned attic. Exception: Dedicated forced-air systems serving only a garage shall not be prohibited from obtaining return air from the garage. 5. Swimming pool enclosures unless they are a dedicated HVAC system.

You Should Know

- Return air is essential to the proper operation of certain HVAC equipment.
- Note that there are seven exceptions to the items specified in Section M1602.2 (above).

CHIMNEYS AND VENTS

TABLE M1803.2 THICKNESS FOR SINGLE-WALL METAL PIPE CONNECTORS

DIAMETER OF CONNECTOR (inches)	GALVANIZED SHEET METAL GAGE NUMBER	MINIMUM THICKNESS (inch)
Less than 6	26	0.019
6 to 10	24	0.024
Over 10 through 16	22	0.029

For SI: 1 inch = 25.4 mm.

TABLE M1803.3.4 CHIMNEY AND VENT CONNECTOR CLEARANCES TO COMBUSTIBLE MATERIALS[a]

TYPE OF CONNECTOR	MINIMUM CLEARANCE (inches)
Single-wall metal pipe connectors: Oil and solid-fuel appliances ❶ Oil appliances listed for use with Type L vents	18 9
Type L vent piping connectors: Oil and solid-fuel appliances Oil appliances listed for use with Type L vents	9 3[b]

For SI: 1 inch = 25.4 mm.
a. These minimum clearances apply to unlisted single-wall chimney and vent connectors. Reduction of required clearances is permitted as in Table M1306.2.
b. When listed Type L vent piping is used, the clearance shall be in accordance with the vent listing.

No.	Code	Description
	M1801.1	Fuel-burning appliances shall be vented to the outdoors in accordance with their listing and label and manufacturer's installation instructions except appliances listed and labeled for unvented use. Venting systems shall consist of approved chimneys or vents or venting assemblies that are integral parts of labeled appliances.
	M1803.2	Chimney and vent connectors for oil- and solid-fuel-burning appliances shall be constructed of factory-built chimney material, Type L vent material, or single-wall metal pipe having resistance to corrosion and heat and thickness not less than that of galvanized steel as specified in Table M1803.2.
❶	M1803.3.4	Connectors shall be installed with clearance to combustibles as set forth in Table M1803.3.4.
	M1804.2.1	Vents passing through a roof shall extend through flashing and terminate in accordance with the manufacturer's installation requirements.

HVAC

CHIMNEYS AND VENTS *(cont.)*		
No.	**Code**	**Description**
	M1804.2.6	Mechanical draft systems shall comply with UL 378 and shall be installed in accordance with their listing, the manufacturer's installation instructions, and, except for direct-vent appliances, the following requirements: 1. The vent terminal shall be located not less than 3′ above a forced-air inlet located within 10′. 2. The vent terminal shall be located not less than 4′ below, 4′ horizontally from, or 1′ above any door, window, or gravity air inlet into a dwelling. 3. The vent termination point shall not be located closer than 3′ to an interior corner formed by two walls perpendicular to each other. 4. The bottom of the vent terminal shall be located at least 12″ above finished ground level. 5. The vent termination shall not be mounted directly above or within 3′ horizontally of an oil tank vent or gas meter. 6. Power exhauster terminations shall be located not less than 10′ from lot lines and adjacent buildings. 7. The discharge shall be directed away from the building.

You Should Know

- This chapter in the IRC code book regulates chimneys serving solid-fuel and oil-burning equipment. Chapter 24 in the IRC code book regulates gas-fired equipment.

PROHIBITED LOCATIONS

No.	Code	Description
❶	G2406.2	Gas appliances shall not be located in sleeping rooms, bathrooms, toilet rooms, storage closets, or surgical rooms, or in a space that opens only into such rooms or spaces, *except* where the installation complies with one of the following: 1. The appliance is a direct-vent appliance installed in accordance with the conditions of the listing and the manufacturer's instructions. 2. Vented room heaters, wall furnaces, vented decorative appliances, vented gas fireplaces, vented gas fireplace heaters, and decorative appliances for installation in vented solid-fuel-burning fireplaces are installed in rooms that meet the required volume criteria of Section G2407.5 in the IRC code book. 3. A single wall-mounted unvented room heater is installed in a bathroom and such unvented room heater is equipped as specified in Section G2445.6 in the IRC code book and has an input rating not greater than 6000 Btu/h. The bathroom shall meet the required volume criteria of Section G2407.5 in the IRC code book. 4. A single wall-mounted *unvented room heater* is installed in a bedroom and such unvented room heater is equipped as specified in Section G2445.6 in the IRC code book and has an input rating not greater than 10,000 Btu/h (2.93 kW). The bedroom shall meet the required volume criteria of Section G2407.5 in the IRC code book. 5. The appliance is installed in a room or space that opens only into a bedroom or bathroom, and such room or space is used for no other purpose and is provided with a solid weather-stripped door equipped with an approved self-closing device. All combustion air shall be taken directly from the outdoors in accordance with Section G2407.6 in the IRC code book. 6. Clothes dryer is installed in a residential bathroom or toilet room having a permanent opening with an area of not less than 100 square inches that communicates with a space outside of a sleeping room, bathroom, toilet room, or storage closet.

INDOOR COMBUSTION AIR

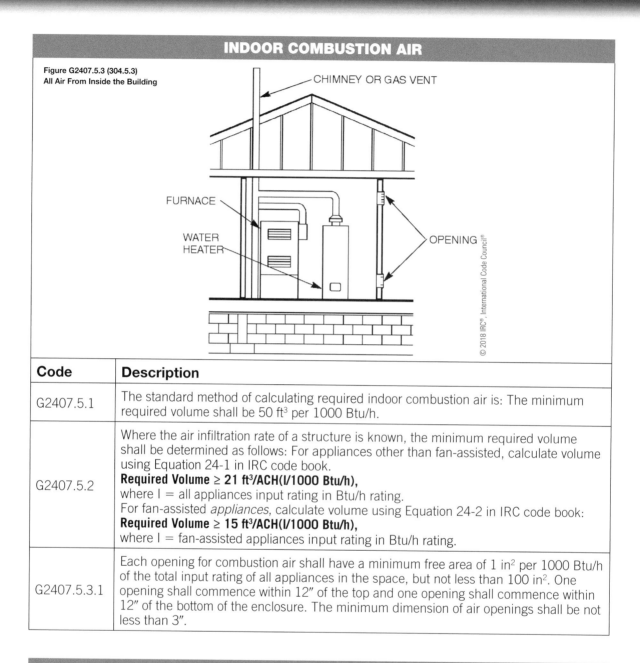

Figure G2407.5.3 (304.5.3)
All Air From Inside the Building

CHIMNEY OR GAS VENT

FURNACE

WATER HEATER

OPENING

© 2018 IRC®, International Code Council®

Code	Description
G2407.5.1	The standard method of calculating required indoor combustion air is: The minimum required volume shall be 50 ft³ per 1000 Btu/h.
G2407.5.2	Where the air infiltration rate of a structure is known, the minimum required volume shall be determined as follows: For appliances other than fan-assisted, calculate volume using Equation 24-1 in IRC code book. **Required Volume ≥ 21 ft³/ACH(I/1000 Btu/h),** where I = all appliances input rating in Btu/h rating. For fan-assisted *appliances*, calculate volume using Equation 24-2 in IRC code book: **Required Volume ≥ 15 ft³/ACH(I/1000 Btu/h),** where I = fan-assisted appliances input rating in Btu/h rating.
G2407.5.3.1	Each opening for combustion air shall have a minimum free area of 1 in² per 1000 Btu/h of the total input rating of all appliances in the space, but not less than 100 in². One opening shall commence within 12″ of the top and one opening shall commence within 12″ of the bottom of the enclosure. The minimum dimension of air openings shall be not less than 3″.

You Should Know

- Where the air infiltration rate of a structure is known, the minimum required volume is calculated through a scientific formula.

OUTDOOR COMBUSTION AIR

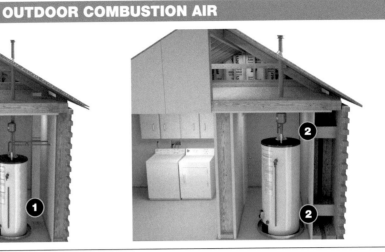

No.	Code	Description
❶	G2407.6	The minimum dimension of air openings shall be not less than 3″.
❷	G2407.6.1	Two permanent openings may be used, one commencing within 12″ of the top and one commencing within 12″ of the bottom of the enclosure, shall be provided. The openings shall communicate directly, or by ducts, with the outdoors, or spaces that freely communicate with the outdoors. • Where directly communicating with the outdoors, or where communicating with the outdoors through vertical ducts, each opening shall have a minimum free area of 1 in^2 per 4000 Btu/h of total input rating of all appliances in the enclosure. • Where communicating with the outdoors through horizontal ducts, each opening shall have a minimum free area of not less than 1 in^2 per 2000 Btu/h of the total input rating of all appliances in the enclosure. One permanent opening may be used, commencing within 12″ of the top of the enclosure, shall be provided. The appliance shall have clearances of at least 1″ from the sides and back and 6″ from the front of the appliance.
	G2407.6.2	One permanent opening, commencing within 12″ of the top of the enclosure, shall be provided. The appliance shall have clearances of at least 1″ from the sides and back and 6″ from the front of the appliance. The opening shall directly communicate with the outdoors or through a vertical or horizontal duct to the outdoors, or spaces that freely communicate with the outdoors (see the following Figure G2407.6.2) and shall have a minimum free area of 1 in^2 per 3000 Btu/h of the total input rating of all appliances located in the enclosure and not less than the sum of the areas of all vent connectors in the space.

Example: Calculate the net opening (in square inches) required for a 65,000 Btu appliance.
Solution: 65,000/4000 = 16.25 in^2
The net area of the openings required for each of the two vents is 16.25 in^2.

OUTDOOR COMBUSTION AIR *(cont.)*

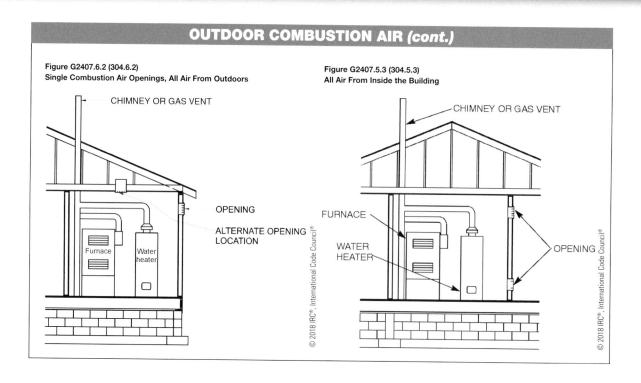

Figure G2407.6.2 (304.6.2)
Single Combustion Air Openings, All Air From Outdoors

CHIMNEY OR GAS VENT

OPENING

ALTERNATE OPENING LOCATION

Furnace

Water heater

© 2018 IRC®, International Code Council®

Figure G2407.5.3 (304.5.3)
All Air From Inside the Building

CHIMNEY OR GAS VENT

FURNACE

WATER HEATER

OPENING

© 2018 IRC®, International Code Council®

COMBINATION INDOOR AND OUTDOOR COMBUSTION AIR

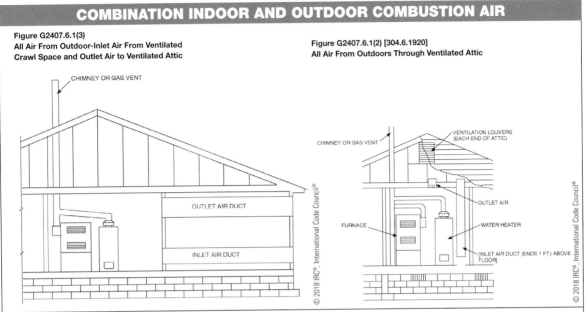

Figure G2407.6.1(3)
All Air From Outdoor-Inlet Air From Ventilated Crawl Space and Outlet Air to Ventilated Attic

Figure G2407.6.1(2) [304.6.1920]
All Air From Outdoors Through Ventilated Attic

Code	Description
	The use of a combination of indoor and outdoor combustion air shall be in accordance with Sections G2407.7.1 through G2407.7.3 in the IRC code book.
G2407.7	Where used, openings connecting the interior spaces shall comply with Section G2407.5.3 in the IRC code book. Outdoor opening(s) shall be located in accordance with Section G2407.6 in the IRC code book. The outdoor opening(s) size shall be calculated in accordance with the following: 1. The ratio of interior spaces shall be the available volume of all communicating spaces divided by the required volume. 2. The outdoor size reduction factor shall be one minus the ratio of interior spaces. 3. The minimum size of outdoor opening(s) shall be the full size of outdoor opening(s) calculated in accordance with Section G2407.6 in the IRC code book, multiplied by the reduction factor. The minimum dimension of air openings shall be not less than 3".

You Should Know

- Engineering design may be used to determine combustion air installations as long as there is an adequate supply of combustion, ventilation, and dilution air.
- Where all combustion air is provided by a mechanical air-supply system, the combustion air shall be supplied from the outdoors at a rate not less than 0.35 cfm per 1000 Btu/h of total input rating of all appliances located within the space.
- Where exhaust fans are installed, makeup air shall be provided to replace the exhausted air.

INSTALLATION OF EQUIPMENT AND PIPING SYSTEM

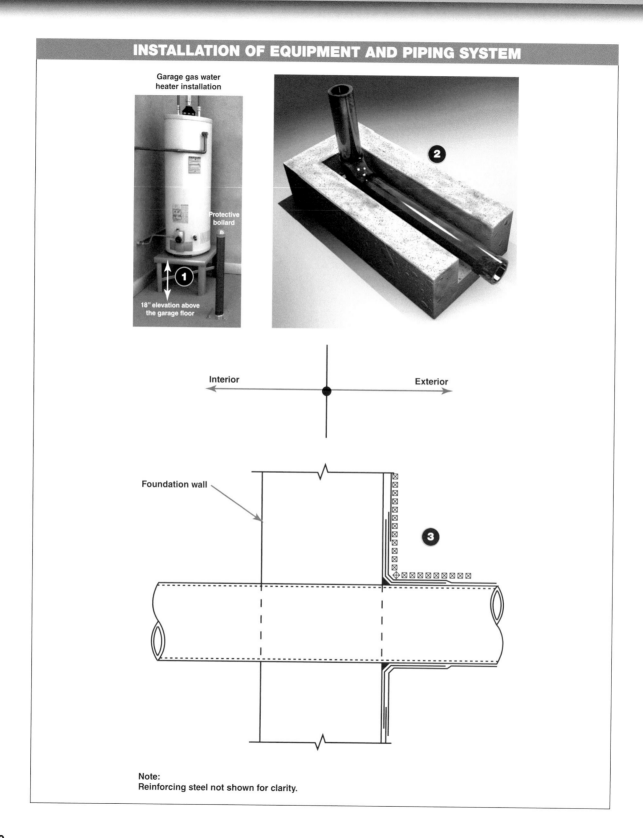

Garage gas water heater installation

Protective bollard

1

18" elevation above the garage floor

2

Interior

Exterior

Foundation wall

3

Note:
Reinforcing steel not shown for clarity.

INSTALLATION OF EQUIPMENT AND PIPING SYSTEM *(cont.)*

No.	Code	Description
❶	G2408.2	Equipment and appliances having an ignition source shall be elevated such that the source of ignition is not less than 18″ above the floor in hazardous locations and public garages, private garages, repair garages, motor fuel-dispensing facilities, and parking garages.
	G2408.4	Equipment and appliances installed at grade level shall be supported on a level concrete slab or other approved material extending not less than 3″ above adjoining grade or shall be suspended not less than 6″ above adjoining grade.
	G2409.1	This section shall govern the reduction in required clearances to combustible materials, including gypsum board and combustible assemblies for chimneys, vents, appliances, devices, and equipment. Drywall has a paper facing that burns and is considered combustible and must meet the clearances indicated in this section.
	G2412.10	Piping, tubing, and fittings shall be manufactured to the applicable referenced standards, specifications, and performance criteria listed in G2414.
	G2415.5	Fittings installed in concealed locations shall be limited to: 1. Threaded elbows, tees and couplings 2. Brazed fittings 3. Welded fittings 4. Fittings listed to ANSI LC1/CSA 6.26 or ANSI LC4/CSA 6.32
❷	G2415.8	Piping in solid floors shall be laid in channels in the floor and covered in a manner that will allow access to the piping with a minimum amount of damage to the building.
❸	G2415.8.1	Conduit shall extend into an occupiable portion of the building and, at the point where the conduit terminates in the building, the space between the conduit and the gas piping shall be sealed to prevent the possible entrance of any gas leakage.
		The conduit shall extend not less than 2″ beyond the point where the pipe emerges from the floor. If the end sealing is capable of withstanding the full pressure of the gas pipe, the conduit shall be designed for the same pressure as the pipe. Such conduit shall extend not less than 4″ outside of the building, shall be vented above grade to the outdoors, and shall be installed to prevent the entrance of water and insects.
	G2415.8.2	Where the conduit originates and terminates within the same building, the conduit shall originate and terminate in an accessible portion of the building and shall not be sealed. The conduit shall extend not less than 2″ beyond the point where the pipe emerges from the floor.

HVAC

CLEARANCES

TABLE G2409.2 (308.2)ᵃ THROUGH k
REDUCTION OF CLEARANCES WITH SPECIFIED FORMS OF PROTECTION

WHERE THE REQUIRED CLEARANCE WITH NO PROTECTION FROM APPLIANCE, VENT CONNECTOR, OR SINGLE-WALL METAL PIPE IS: (inches)

Allowable clearances with specified protection (inches)

Use Column 1 for clearances above appliance or horizontal connector. Use Column 2 for clearances from appliance, vertical connector, and single-wall metal pipe.

TYPE OF PROTECTION APPLIED TO AND COVERING ALL SURFACES OF COMBUSTIBLE MATERIAL WITHIN THE DISTANCE SPECIFIED AS THE REQUIRED CLEARANCE WITH NO PROTECTION [see Figures G2409.2(1), G2409.2(2), and G2409.2(3)]	36		18		12		9		6	
	Above Col. 1	Sides and rear Col. 2	Above Col. 1	Sides and rear Col. 2	Above Col. 1	Sides and rear Col. 2	Above Col. 1	Sides and rear Col. 2	Above Col. 1	Sides and rear Col. 2
1. 3½-inch-thick masonry wall without ventilated airspace	—	24	—	12	—	9	—	6	—	5
2. ½-inch insulation board over 1-inch glass fiber or mineral wool batts	24	18	12	9	9	6	6	5	4	3
3. 0.024-inch (nominal 24 gage) sheet metal over 1-inch glass fiber or mineral wool batts reinforced with wire on rear face with ventilated airspace	18	12	9	6	6	4	5	3	3	3
4. 3½-inch-thick masonry wall with ventilated airspace	—	12	—	6	—	6	—	6	—	6
5. 0.024-inch (nominal 24 gage) sheet metal with ventilated airspace	18	12	9	6	6	4	5	3	3	2
6. ½-inch-thick insulation board with ventilated airspace	18	12	9	6	6	4	5	3	3	3
7. 0.024-inch (nominal 24 gage) sheet metal with ventilated airspace over 0.024-inch (nominal 24 gage) sheet metal with ventilated airspace	18	12	9	6	6	4	5	3	3	3
8. 1-inch glass fiber or mineral wool batts sandwiched between two sheets 0.024-inch (nominal 24 gage) sheet metal with ventilated airspace	18	12	9	6	6	4	5	3	3	3

For SI: 1 inch = 25.4 mm, °C = [(°F -32)/1.8], 1 pound per cubic foot = 16.02 kg/m³, 1 Btu per inch per square foot per hour per °F = 0.144 W/m² K.

a. Reduction of clearances from combustible materials shall not interfere with combustion air, draft hood clearance and relief, and accessibility of servicing.
b. All clearances shall be measured from the outer surface of the combustible material to the nearest point on the surface of the appliance, disregarding any intervening protection applied to the combustible material.
c. Spacers and ties shall be of noncombustible material. No spacer or tie shall be used directly opposite an appliance or connector.
d. For all clearance reduction systems using a ventilated airspace, adequate provision for air circulation shall be provided [see Figures G2409.2(2) and G2409.2(3)].
e. There shall be at least 1 inch between clearance reduction systems and combustible walls and ceilings for reduction systems using ventilated airspace.
f. Where a wall protector is mounted on a single flat wall away from corners, it shall have a minimum 1-inch air gap. To provide air circulation, the bottom and top edges, or only the side and top edges, or all edges shall be left open.
g. Mineral wool batts (blanket or board) shall have a minimum density of 8 pounds per cubic foot and a minimum melting point of 1,500°F.
h. Insulation material used as part of a clearance reduction system shall have a thermal conductivity of 1.0 Btu per inch per square foot per hour per °F or less.
i. There shall be at least 1 inch between the appliance and the protector. In no case shall the clearance between the appliance and the combustible surface be reduced below that allowed in this table.
j. All clearances and thicknesses are minimum; larger clearances and thicknesses are acceptable.
k. Listed single-wall connectors shall be installed in accordance with the manufacturer's installation instructions.

No.	Code	Description
1	G2408.5	Heat-producing equipment and appliances shall be installed to maintain the required clearances to combustible construction as specified in the listing and manufacturer's instructions.
2	G2409.2	The allowable clearance reduction shall be based on one of the methods specified in Table G2409.2 or shall utilize an assembly listed for such application.

Table title: **CLEARANCES *(cont.)***

HVAC

ELECTRICAL BONDING

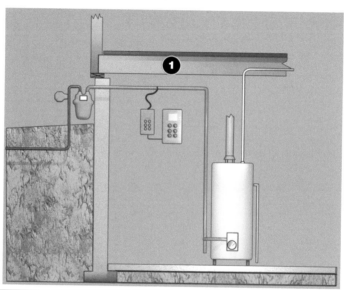

No.	Code	Description
1	G2411.1	Each aboveground portion of a gas piping system other than corrugated stainless-steel tubing (CSST) that is likely to become energized shall be electrically continuous and bonded to an effective ground-fault current path. CSST gas piping systems shall be bonded to the electrical-service grounding electrode system. The bonding jumper shall connect to metallic pipe or fitting between the point of delivery and the first downstream CSST fitting. The bonding jumper shall be not smaller than 6 AWG copper wire or equivalent.
	G2411.2 G2411.2.3	These sections deal with CSST not listed as arc-resistant jacket. The bonding jumper shall connect to a metallic pipe, pipe fitting, or CSST fitting. The bonding jumper shall not exceed 75′ in length.
	G2411.3	Arc-resistant CSST must be listed with ANSI LC1/CSA 6.26. The CSST shall be electrically continuous and bonded to an effective ground fault current path.

PIPE SIZING

$$D = \frac{Q^{0.381}}{19.17\left(\dfrac{\Delta H}{C_r \times L}\right)^{0.206}}$$ **(Equation 24-3)**

where:
D = Inside diameter of *pipe*, inches (mm).
Q = Input rate *appliance(s)*, cubic feet per hour at 60°F (16°C) and 30-inch mercury column.
P_1 = Upstream pressure, psia (P_1 + 14.7).
P_2 = Downstream pressure, psia (P_2 + 14.7).
L = Equivalent length of *pipe*, feet.
ΔH = *Pressure drop*, inch water column (27.7 inch water column = 1 psi).

❷

$$D = \frac{Q^{0.381}}{18.93\left[\dfrac{(p_1^2 - p_2^2)\times Y}{C_r \times L}\right]^{0.206}}$$ **(Equation 24-4)**

where:
D = Inside diameter of *pipe*, inches (mm).
Q = Input rate *appliance(s)*, cubic feet per hour at 60°F (16°C) and 30-inch mercury column.
P_1 = Upstream pressure, psia (P_1 + 14.7).
P_2 = Downstream pressure, psia (P_2 + 14.7).
L = Equivalent length of *pipe*, feet.
ΔH = *Pressure drop*, inch water column (27.7 inch water column = 1 psi).

No.	Code	Description
❶	G2413.3	Gas piping shall be sized in accordance with one of the following: 1. Pipe sizing tables or sizing equations in accordance with Section G2413.4 or G2413.5. 2. The sizing tables included in a listed piping system's manufacturer's installation instructions. 3. Other approved engineering methods.
	G2413.4.1	The pipe size of each section of gas piping shall be determined using the longest length of piping from the point of delivery to the most remote outlet and the load of the section.
❷	Equations 24-3 and 24-4	Low- and high-pressure gas pipe sizing can be designed using the two equations, 24-3 and 24-4. Normally, this would be an engineered design.

You Should Know

- Pipe size is based on type of gas and piping, gas pressure throughout the system, and input rate for appliances.
- The following several pages walk you through the most common method.

LONGEST-PIPE-LENGTH METHOD

Pipe size is based on several criteria including the type of pipe material, gas capacity of each appliance, the distance from the meter to the appliance, and the pressure throughout the gas system.

The IRC code sets out a performance standard in Section G2413.1:

"Piping systems shall be of such size and so installed as to provide a supply of gas sufficient to meet the maximum *demand* and supply gas to each *appliance* inlet at not less than the minimum supply pressure required by the *appliance*."

Let's take a closer look at these variables:

- Btu Demand is Based on the Appliance Manufacturer: This measurement of gas demand is from the manufacturer and normally on a visible plate on the appliance. Where the input is not available for a particular appliance, Table G2413.2 offers approximate values.
- Capacity of Cubic Feet of Gas/Hour: This is a volume of gas needed to be delivered to the appliance for its effective operation. You will need to convert Btu/h to this factor. To determine this, divide Btu/h by 1000.
- Material Type: There are several materials suitable for gas supply. These include steel, steel tubing, copper tubing, CSST, and plastic pipe. Cast iron shall not be used as gas pipe.
- Length to Farthest Appliance in Feet: This is a measure from the gas meter to the farthest gas appliance. This includes vertical measurement up and down a wall.
- Specific Gravity: As applied to gas, specific gravity is the ratio of the weight of a given volume to that of the same volume of air, both measured under the same condition.
- Pressure Drop: The loss in pressure due to friction or obstruction in pipes, valves, fittings, regulators, and burners.

The total connected demand is used as the basis for pipe sizing. You must assume that *all* appliances are drawing gas at the maximum possible demand. Each appliance will (or should have) a metal plate that indicates its Btu/h demand. In the upper right-hand corner of each table is a chart that defines the parameters of the table. Select the appropriate table based on the type of pipe material, the type of gas (natural or LP), the gas pressure, pressure drop, and specific gravity.

Essentially, you use the appropriate table based on the parameters (gas type, inlet pressure, pressure drop, and specific gravity), then use the equivalent length in feet along with the calculated Btu/h or capacity in cubic feet of gas per hour to determine the pipe size in diameter.

Let's try a few...

For this example, consider using CSST, natural gas, inlet pressure less than 2 psi, pressure drop ½ inch water column, with a specific gravity of 0.60. This will lead you to Table G2413.4(5). For this example, let's assume there is only one appliance (gas forced-air heating system) with a 140,000 Btu/h rating. The length of pipe is 75' from the gas meter to the appliance.

LONGEST-PIPE-LENGTH METHOD *(cont.)*

	Gas	Natural
	Inlet Pressure	Less than 2 psi
	Pressure Drop	0.5 in. w.c.
	Specific Gravity	0.60

TABLE G2413.4(5) [402.4(15)] CORRUGATED STAINLESS STEEL TUBING (CSST)

Flow Desig-nation	13	15	18	19	23	25	30	31	37	39	46	48	60	62
Length (ft)	Capacity in Cubic Feet of Gas per Hour													
5	46	63	115	134	225	270	471	546	895	1,037	1,790	2,070	3,660	4,140
10	32	44	82	95	161	192	330	383	639	746	1,260	1,470	2,600	2,930
15	25	35	66	77	132	157	267	310	524	615	1,030	1,200	2,140	2,400
20	22	31	58	67	116	137	231	269	456	536	888	1,050	1,850	2,080
25	19	27	52	60	104	122	206	240	409	482	793	936	1,660	1,860
30	18	25	47	55	96	112	188	218	374	442	723	856	1,520	1,700
40	15	21	41	47	83	97	162	188	325	386	625	742	1,320	1,470
50	13	19	37	42	75	87	144	168	292	347	559	665	1,180	1,320
60	12	17	34	38	68	80	131	153	267	318	509	608	1,080	1,200
70	11	16	31	36	63	74	121	141	248	295	471	563	1,000	1,110
80	10	15	29	33	60	69	113	132	232	277	440	527	940	1,040
90	10	14	28	32	57	65	107	125	219	262	415	498	887	983
100	9	13	26	30	54	62	101	118	208	249	393	472	843	933
150	7	10	20	23	42	48	78	91	171	205	320	387	691	762
200	6	9	18	21	38	44	71	82	148	179	277	336	600	661
250	5	8	16	19	34	39	63	74	133	161	247	301	538	591
300	5	7	15	17	32	36	57	67	95	148	226	275	492	540

TUBE SIZE (EHD)

For SI: 1 inch = 25.4 mm, 1 foot = 304.8 mm, 1 pound per square inch = 6.895 kPa, 1-inch water column = 0.2488 kPa, 1 British thermal unit per hour = 0.2931 W, 1 cubic foot per hour = 0.0283 m³/h, 1 degree = 0.01745 rad.

© 2018 IRC®, International Code Council®

Notes:
1. Table includes losses for four 90-degree bends and two end fittings. Tubing runs with larger numbers of bends and/or fittings shall be increased by an equivalent length of tubing to the following equation: $L = 1.3n$, where L is additional length (feet) of tubing and n is the number of additional fittings and/or bends.
2. EHD—Equivalent Hydraulic Diameter, which is a measure of the relative hydraulic efficiency between different tubing sizes. The greater the value of EHD, the greater the gas capacity of the tubing.
3. All table entries have been rounded to three significant digits.

Table G2413.4(5) refers to maximum capacity in cubic feet of gas per hour. If you don't have this but have the Btu/h ratings on all the appliances, you must convert them. Do this by adding all the Btu/h ratings, then dividing that figure by 1000. Divide the 140,000 Btu/h by 1000 and get: 140 ft³ of gas per hour.

Now enter the table in the flow designation column and scroll down to find 75′. Since there is no entry for that length, use the 80′ row. Then scroll over until you find 140 (cubic feet per hour). You find 132, which is too small. Use the next column, which is 232, which is larger than 140. Move up that column to find the necessary pipe size of 37 EHD (equivalent hydraulic diameter).

LONGEST-PIPE-LENGTH METHOD *(cont.)*

The required CSST size is 37 EHD for this application.

Let's begin by determining the longest length in a gas pipe distribution system. Calculating the longest length is a matter of adding the lengths from the point of delivery (gas meter) to each appliance. Then use the longest distance for calculating the required pipe size in each table. Let's try one:

12 + 22 + 14 = 4'
12 + 22 + 12 + 13 = 59'
12 + 22 + 12 + 12 = 70'
12 + 22 + 12 + 12 + 18 = 76' (This is the longest.)

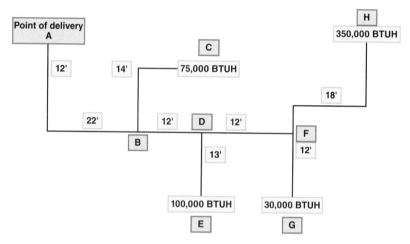

Step 1. Select the proper table in Chapter 24 in the IRC code book based on type of gas, inlet pressure, pressure drop, and specific gravity. These variables are listed in a section on the upper right side of each table.

Step 2. The longest length is also called the flow designation length in each table. You calculate this by separately adding the length of each branch, one at a time. In the example above, the distance to the farthest appliance is 76'.

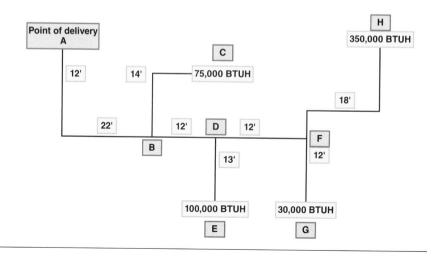

LONGEST-PIPE-LENGTH METHOD *(cont.)*

Step 1. When you have determined the longest length, in the example, go to the proper table in the IRC code book.

Step 2. Enter the column for length and find the appropriate row (at least equal to 76′). In this case it is 80′.

Step 3. This is the row to use when sizing the pipe's diameter.

Let's try another task using the same example. In the figure illustrated, calculate the pipe in the last section from F to H.

1. Begin by determining the cubic feet per hour. Do this by dividing Btu demand:

   75,000 + 100,000 + 350,000 + 30,000 = 555,000 Btu/h.

2. Now convert to cubic feet of gas per hour: 555,000/1000 = 555 cubic feet/hour

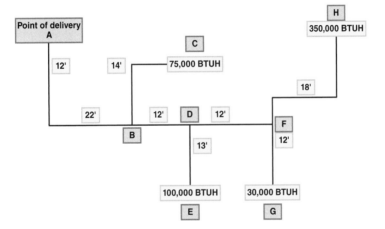

3. Next use the capacity of the cubic feet of gas per hour for the fixture—H: 350,000/1000 = 350 cubic feet per hour. Then, using the longest distance, 76′ (between A and H), enter the table to determine pipe size. For CSST, that is Table G2413.4(5) in the IRC code book. Go down to 80′ on the first column, then scroll over on that row to find at least 350 cubic feet/hour. You must use the number above that or 440. Go up that column to read the required pipe size. The size is 46 EHD.

4. But what if the gas pipe was to be polyethylene plastic pipe with the same variables for the gas. That is natural gas, inlet pressure less than 2 psi, pressure drop 0.5 inch water column, and a specific gravity of 0.5. Select Table G2413.4(7) in the IRC code book and use the 80′ length in the first column. Scroll over until you find a capacity equal or greater than 350. You then use the capacity of 409. Go up that column to determine that the pipe size of 1¼″ and a designation SDR 10.00. Next, let's determine the required length of each segment.

Given: The gas variables are natural gas, inlet pressure less than 2 psi, pressure drop 0.5 inch water column, and a specific gravity of 0.5.

Pipe materials will be polyethylene plastic pipe. Therefore, use Table G2413.4(7) in the IRC code book. You can start at either end but you will always use the longest length for each pipe segment. In this case that is 76′.

LONGEST-PIPE-LENGTH METHOD *(cont.)*

Step 1. To find pipe segment FG, use the capacity at G (30) and enter the table at 80' length and scroll over until you pass 30. Find 65 on the second column and scroll up to find ½"-diameter pipe.

Step 2. To find pipe segment FH, use the capacity at H (350) and enter the table at 80' length and scroll over until you pass 350. Find 409 on the column and scroll up to find and use 1¼".

Step 3. To find pipe segment DF, use the capacity at G and H (380) and enter the table at 80' length and scroll over until you pass 380. Find 409 on the column and scroll up to find and use 1¼".

Step 4. To find pipe segment DE, use the capacity at E (100) and enter the table at 80' length and scroll over until you pass 100. Find 131 on the column and scroll up to find and use ¾".

Step 5. To find pipe segment BD, use the capacity at E, G, and H (480) and enter the table at 80' length and scroll over until you pass 480. Find 617 on the column and scroll up to find and use 1½".

Step 5. To find pipe segment BC, use the capacity at C (75) and enter the table at 80' length and scroll over until you pass 75. Find 131 on the column and scroll up to find and use ¾".

Step 6. To find pipe segment AB, use the capacity at C, E, G, and H (555) and enter the table at 80' length and scroll over until you pass 555. Find 617 on the column and scroll up to find and use 1½".

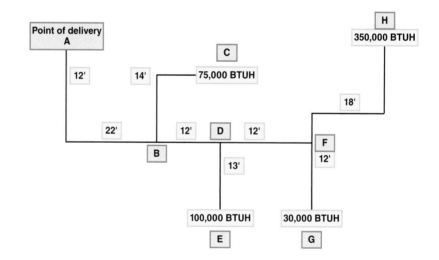

AB–1½"
BC–¾"
BD–1½"
DE–¾"
DF–1¼"
FH–1¼"
FG–½"
Note: Always use the longest length when sizing branch pipe lines.

PIPE MATERIALS, INSPECTION, TESTING, AND PURGING

No.	Code	Description
1	G2414.4.1	Cast-iron pipe shall not be used.
2	G2414.4.2	Steel, stainless steel, and wrought iron shall not be lighter than schedule 10 and shall comply with ASME B36.10, 10M or 1. ASTM A53/A53M 2. ASTM A106 3. ASTM A312
3	G2417.1.1	Inspection shall consist of visual examination, during or after manufacture, fabrication, assembly, or pressure tests as appropriate.
4	G2417.2	The test medium shall be air, nitrogen, carbon dioxide, or inert gas. Oxygen shall *not* be used.

PIPE MATERIALS, INSPECTION, TESTING, AND PURGING *(cont.)*

No.	Code	Description
5	G2417.4.1	The test pressure to be used shall not be less than 1½ times the proposed maximum working pressure but not less than 3 psig, irrespective of design pressure.
6	G2417.4.2	The test duration shall be 10 minutes.
7	G2417.6.2	During the process of turning gas on to the system of new gas piping, the entire system shall be inspected to determine that there are no open fittings or ends and that all valves at unused outlets are closed, plugged, or capped.

You Should Know

- Copper and brass tubing shall not be used if the gas contains more than an average of 0.3 grains of hydrogen sulfide per 100 standard cubic feet of gas.
- There are a variety of material standards that must be met for each material type. For example, steel and wrought iron pipe must be at least Schedule 40 and comply with either ASTM B36.10, 10M; ASTM A53/A 53M; ASTM A312, or ASTM A106. Copper tubing must comply with ASTM B88 or ASTM B280.
- Plastic pipe, tubing, and fittings must comply with ASTM D2513 and be marked "Gas" and "ASTM D2513" per Section G2414.6 in the IRC code book.

GAS SHUTOFF VALVES AND APPLIANCE CONNECTORS

No.	Code	Description
1	G2420.5.1	The shutoff valve shall be located in the same room as the appliance. The shutoff valve shall be within 6′ of the appliance and shall be installed upstream of the union, connector, or quick disconnect device it serves. Such shutoff valves shall be provided with access. Shutoff valves installed behind movable appliances shall be considered to provide access where installed behind such appliances.
	G2420.5.2	Shutoff valves for vented decorative appliances, room heaters, and decorative appliances for installation in vented fireplaces shall be permitted to be installed in an area remote from the appliances where such valves are provided with ready access. Such valves shall be permanently identified and shall serve no other appliance. The piping from the shutoff valve shall reach to within 6′ of the appliance.
	G2422.1.2.1	Appliance connectors shall not exceed 6′ in overall length. Measurement shall be made along the centerline of the connector. Only one connector shall be used for each appliance.

PIPING SUPPORT INTERVALS

TABLE G2424.1 SUPPORT OF PIPING

STEEL PIPE, NOMINAL SIZE OF PIPE (inches)	SPACING OF SUPPORTS (feet)	NOMINAL SIZE OF TUBING SMOOTH-WALL (inch O.D.)	SPACING OF SUPPORTS (feet)
½	6	½	4
¾ or 1	8	$\frac{5}{8}$ or ¾	6
1¼ or larger (horizontal)	10	$\frac{7}{8}$ or 1 (horizontal)	8
1¼ or larger (vertical)	Every floor level	1 or larger (vertical)	Every floor level

For SI: 1 inch = 25.4 mm, 1 foot = 304.8 mm.

© 2018 IRC®, International Code Council®

No.	Code	Description
1	G2424.1	Piping shall be supported at intervals not exceeding the spacing specified in Table G2424.1. Spacing of supports for CSST shall be in accordance with the CSST manufacturer's instructions.

GAS VENTS

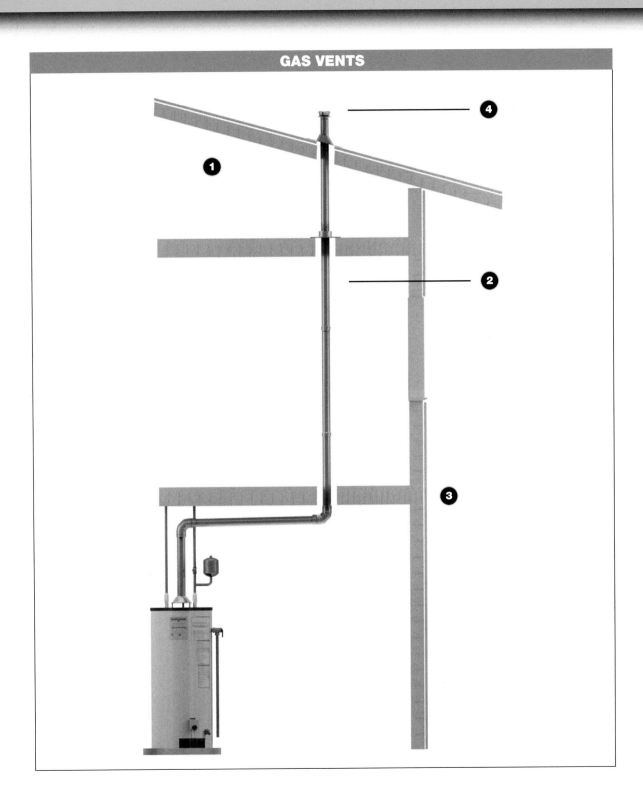

GAS VENTS (cont.)

No.	Code	Description
❶	G2426.3	The application of vents shall be in accordance with Table G2427.4.
❷	G2426.5	Vent systems shall be sized, installed, and terminated in accordance with the vent and appliance manufacturer's installation instructions.
❸	G2427.2	Except as permitted in the IRC code book in Sections G2427.2.1 Direct-vent Appliances, G2427.2.2 Appliances with Integral Vents, and G2425.8.
❹	G2427.6.5	A Type B or L gas vent shall terminate at least 5' in vertical height above the highest connected appliance draft hood or flue collar. A Type B-W gas vent shall terminate at least 12' in vertical height above the bottom of the wall furnace. This applies to many appliances, including a natural-draft gas water heater.
	G2428.2.3	Single-appliance venting configurations with zero (0) lateral lengths in Tables G2428.2(1) and G2428.2(2) in the IRC code book shall not have elbows in the venting system. Single-appliance venting configurations with lateral lengths include two 90-degree elbows.
	Table G2428.2	The most common vent for a gas water heater is a Type B (double-wall) vent. This vent is determined from using the tables explained by the examples on the following pages.

YOU SHOULD KNOW

- Appliances not required to be vented include direct-vent appliances, appliances with integral vents, ranges, built-in domestic cooking units listed for optional venting, hot plates, and laundry stoves, Type I clothes dryers, refrigerators, counter appliances, and room heaters listed for unvented use.
- Everything else must be adequately vented.
- Appliance and equipment vent terminals shall be located so that doors cannot swing within 12" horizontally of the vent terminal. Door stops or closures shall not be installed to obtain this clearance.

SIZING OF APPLIANCE VENTING SYSTEMS

① TABLE G2428.2(1) [504.2(1)] - TYPE B

② Height (H) (feet)	③ Lateral (L) (feet)	⑥ 3			4			5		
		⑤ FAN		NAT	FAN		NAT	FAN		NAT
		Min.	Max.	Max.	Min.	Max.	Max.	Min.	Max.	Max.
6	0	0	78	46	0	152	86	0	251	141
	2	13	51	36	18	97	67	27	157	105
	4	21	49	34	30	94	64	39	153	103
	6	④ 25	46	32	36	91	61	47	149	100
8	0	0	84	50	0	165	94	0	276	
	2	12	57	40	16	109	75	25	178	
	5	23	53	38	32	103	71	42	171	
	8	28	49	35	39	98	66	51	164	

APPLIANCE INPUT R

No.	Code	Description
①	Tables G2428.2(1) and (2)	Select the appropriate Table G2428.2(1) or (2) whether (1) the appliance is connected directly to the vent or (2) there is a single-wall metal connector. A Type B vent is typical for gas-fueled water heaters. "Connected to vent" means the vent extends from the appliance to the vent termination. "Single-wall metal connector" indicates a single-wall metal vent connector extends from the appliance to the vent, commonly through the ceiling.
②	G2428.2	Determine the total vertical rise from the flue collar to the vent termination in feet. Where the vent size determined from the tables is smaller than the appliance draft hood outlet or flue collar, the smaller size shall be permitted to be used provided all of the following are met: 1. The total vent height (H) is at least 10′. 2. Vents for appliance draft hood outlets or flue collars 12″ in diameter or smaller are not reduced more than one table size. 3. Vents for appliance draft hood outlets or flue collars larger than 12″ in diameter are not reduced more than two table sizes. 4. The maximum capacity listed in the tables for a fan-assisted appliance is reduced by 10% by maximum table capacity. 5. The draft hood outlet is greater than 4″ in diameter. Do not connect a 3″-diameter vent to a 4″-diameter draft hood outlet. This provision shall not apply to fan-assisted appliances.
③		Determine any lateral length for the vent. How much lateral length in feet does the vent have? It could be 0.
④	Tables G2428.2(1) and (2)	Establish the cubic feet of gas input of the appliance.
⑤		Choose the type of ventilation for the condition. Water heaters are most commonly natural draft. The other option is fan-assisted ventilation.
⑥		Find the minimum diameter of the vent to be used.

HVAC

VENT TERMINATION

Figure G2427.6.3 (503.6.4)
Gas Vent Termination Locations for Listed Caps 12 Inches or less in Size at least 8 Feet from a Vertical Wall

ROOF SLOPE	H (minimum) FT
Flat to 6/12	1.0
Over 6/12 to 7/12	1.25
Over7/12 to 8/12	1.5
Over 8/12 to 9/12	2.0
Over 9/12 to 10/12	2.5
Over 10/12 to 11/12	3.25
Over 11/12 to 12/12	4.0
Over 12/12 to 14/12	5.0
Over 14/12 to 16/12	6.0
Over 16/12 to 18/12	7.0
Over 18/12 to 20/12	7.5
Over 20/12 to 21/12	8.0

For SI: 1 foot = 304.8 mm.

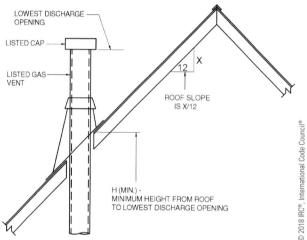

Code	Description
G2427.6.4	A gas vent shall terminate in accordance with one of the following: 1. Gas vents that are 12″ or less in size and located not less than 8′ from a vertical wall or similar obstruction shall terminate above the roof in accordance with Figure G2427.6.3. 2. Gas vents that are over 12″ in size or are located less than 8′ from a vertical wall or similar obstruction shall terminate not less than 2′ above the highest point where they pass through the roof and not less than 2′ above any portion of a building within 10′ horizontally. 3. As provided for direct-vent systems in Section G2427.2.1 in the IRC code book. 4. As provided for appliances with integral vents in Section G2427.2.2 in the IRC code book. 5. As provided for mechanical draft systems in Section G2427.3.3 in the IRC code book.

VENT TERMINATION *(cont.)*

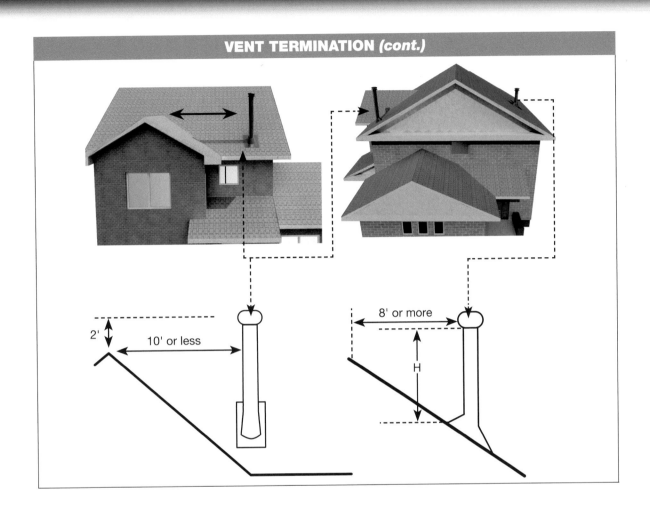

HVAC

VENT CONNECTORS FOR APPLIANCES

TABLE G2427.10.5 (503.10.5)ᵃ CLEARANCES FOR CONNECTORS ❶

APPLIANCE	MINIMUM DISTANCE FROM COMBUSTIBLE MATERIAL			
	Listed Type B gas vent material	Listed Type L vent material	Single-wall metal pipe	Factory-built chimney sections
Listed appliances with draft hoods and appliances listed for use with Type B gas vents	As listed	As listed	6 inches	As listed
Residential boilers and furnaces with listed gas conversion burner and with draft hood	6 inches	6 inches	9 inches	As listed
Residential appliances listed for use with Type L vents	Not permitted	As listed	9 inches	As listed
Listed gas-fired toilets	Not permitted	As listed	As listed	As listed
Unlisted residential appliances with draft hood	Not permitted	6 inches	9 inches	As listed
Residential and low-heat appliances other than above	Not permitted	9 inches	18 inches	As listed
Medium-heat appliances	Not permitted	Not permitted	36 inches	As listed

For SI: 1 inch = 25.4 mm.

a. These clearances shall apply unless the manufacturer's installation instructions for a listed appliance or connector specify different clearances, in which case the listed clearances shall apply.

© 2018 IRC®, International Code Council®

VENT CONNECTORS FOR APPLIANCES *(cont.)*

No.	Code	Description
❶	G2427.10.5	Minimum clearances from vent connectors to combustible material shall be in accordance with Table G2427.10.5.
❷	G2427.10.8	A vent connector shall be as short as practical and the appliance located as close as practical to the chimney or vent.
		The maximum horizontal length of a single-wall connector shall be 75% of the height of the chimney or vent unless engineered.
		The maximum horizontal length of a Type B double-wall connector shall be 100% of the height of the chimney or vent unless engineered.
	G2427.10.11	The entire length of a vent connector shall be provided with ready access for inspection, cleaning, and replacement.

You Should Know

- A vent connector shall be used to connect an appliance to a gas vent, chimney, or single-wall metal pipe, except where the gas vent, chimney, or single-wall metal pipe is directly connected to the appliance.

HVAC

FORCED-AIR WARM-AIR FURNACES

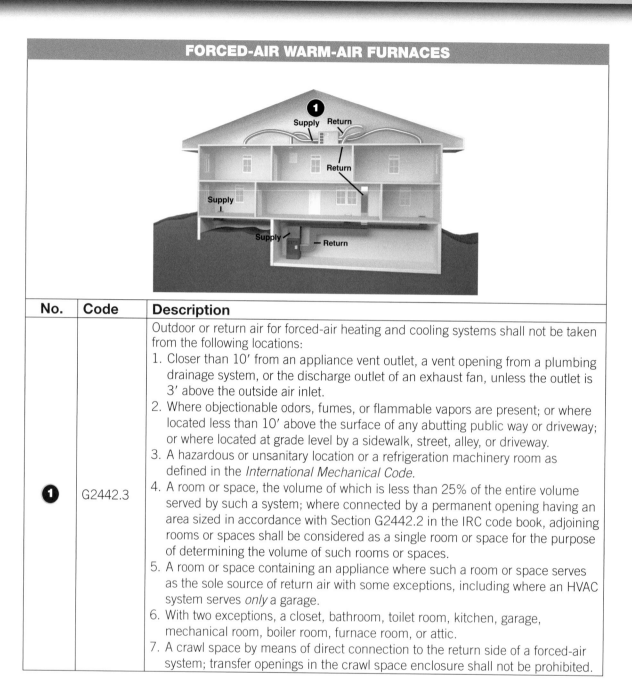

No.	Code	Description
❶	G2442.3	Outdoor or return air for forced-air heating and cooling systems shall not be taken from the following locations: 1. Closer than 10' from an appliance vent outlet, a vent opening from a plumbing drainage system, or the discharge outlet of an exhaust fan, unless the outlet is 3' above the outside air inlet. 2. Where objectionable odors, fumes, or flammable vapors are present; or where located less than 10' above the surface of any abutting public way or driveway; or where located at grade level by a sidewalk, street, alley, or driveway. 3. A hazardous or unsanitary location or a refrigeration machinery room as defined in the *International Mechanical Code.* 4. A room or space, the volume of which is less than 25% of the entire volume served by such a system; where connected by a permanent opening having an area sized in accordance with Section G2442.2 in the IRC code book, adjoining rooms or spaces shall be considered as a single room or space for the purpose of determining the volume of such rooms or spaces. 5. A room or space containing an appliance where such a room or space serves as the sole source of return air with some exceptions, including where an HVAC system serves *only* a garage. 6. With two exceptions, a closet, bathroom, toilet room, kitchen, garage, mechanical room, boiler room, furnace room, or attic. 7. A crawl space by means of direct connection to the return side of a forced-air system; transfer openings in the crawl space enclosure shall not be prohibited.

You Should Know

- Return air from one dwelling unit cannot be discharged into another dwelling unit per Section M1602 IRC.

BOILERS

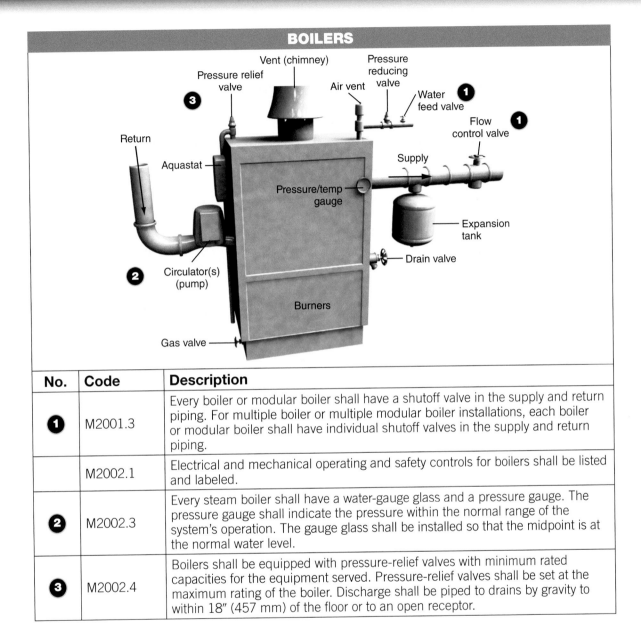

No.	Code	Description
❶	M2001.3	Every boiler or modular boiler shall have a shutoff valve in the supply and return piping. For multiple boiler or multiple modular boiler installations, each boiler or modular boiler shall have individual shutoff valves in the supply and return piping.
	M2002.1	Electrical and mechanical operating and safety controls for boilers shall be listed and labeled.
❷	M2002.3	Every steam boiler shall have a water-gauge glass and a pressure gauge. The pressure gauge shall indicate the pressure within the normal range of the system's operation. The gauge glass shall be installed so that the midpoint is at the normal water level.
❸	M2002.4	Boilers shall be equipped with pressure-relief valves with minimum rated capacities for the equipment served. Pressure-relief valves shall be set at the maximum rating of the boiler. Discharge shall be piped to drains by gravity to within 18″ (457 mm) of the floor or to an open receptor.

HVAC

BOILERS *(cont.)*

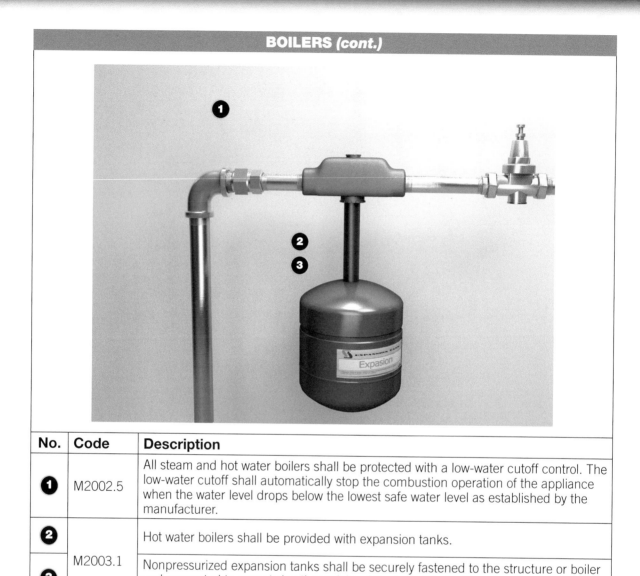

No.	Code	Description
❶	M2002.5	All steam and hot water boilers shall be protected with a low-water cutoff control. The low-water cutoff shall automatically stop the combustion operation of the appliance when the water level drops below the lowest safe water level as established by the manufacturer.
❷	M2003.1	Hot water boilers shall be provided with expansion tanks.
❸		Nonpressurized expansion tanks shall be securely fastened to the structure or boiler and supported to carry twice the weight of the tank filled with water. Provisions shall be made for draining nonpressurized tanks without emptying the system.
	M2002.5	Steam and hot water boiler shall be protected with a low-water cutoff control.

ELECTRICAL CONSIDERATIONS

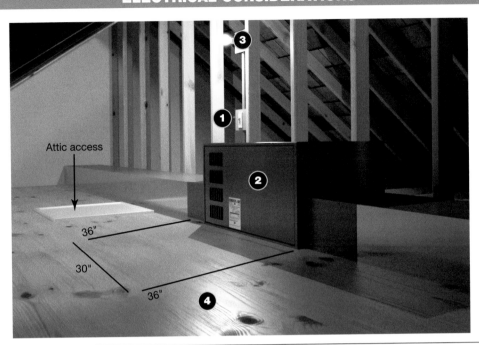

Attic access

36"

30"

36"

No.	Code	Description
1	M1305.1.3.3	For appliances in attics or under floors, a luminaire controlled by a switch located at the required passageway opening and a receptacle outlet shall be installed at or near the appliance location in accordance with Chapter 39 in the IRC code book. Exposed lamps shall be protected from damage by location or lamp guards.
	E3702.11	The ampacity of the conductors supplying multimotor and combination load equipment shall not be less than the minimum circuit ampacity marked on the equipment. The branch-circuit overcurrent device rating shall be the size and type marked on the appliance.
2	E3901.12	A 125-volt, single-phase, 15- or 20-ampere-rated receptacle outlet shall be installed at an accessible location for the servicing of heating, air-conditioning, and refrigeration equipment. The receptacle shall be located on the same level and within 25′ of the heating, air-conditioning, and refrigeration equipment. The receptacle outlet shall not be connected to the load side of the HVAC equipment disconnecting means.
3	E3903.4	In attics, under-floor spaces, utility rooms, and basements, at least one lighting outlet shall be installed where these spaces are used for storage or contain equipment requiring servicing. Such lighting outlet shall be controlled by a wall switch or shall have an integral switch. At least one point of control shall be at the usual point of entry to these spaces. The lighting outlet shall be provided at or near the equipment requiring servicing.

ELECTRICAL CONSIDERATIONS *(cont.)*

No.	Code	Description
4	E3405.2	The dimension of the working space in the direction of access to panelboards and live parts likely to require examination, adjustment, servicing, or maintenance while energized shall be not less than 36″ in depth. Distances shall be measured from the energized parts where such parts are exposed or from the enclosure front or opening where such parts are enclosed. In addition to the 36″ dimension, the work space shall not be less than 30″ wide in front of the electrical equipment and not less than the width of such equipment. In all cases, the work space shall allow at least a 90° opening of equipment doors or hinged panels. Equipment associated with the electrical installation located above or below the electrical equipment shall be permitted to extend not more than 6″ beyond the front of the electrical equipment.

❶ TABLE E3908.12 EQUIPMENT GROUNDING CONDUCTOR SIZING

RATING OR SETTING OF AUTOMATIC OVERCURRENT DEVICE IN CIRCUIT AHEAD OF EQUIPMENT, CONDUIT, ETC., NOT EXCEEDING THE FOLLOWING RATINGS (amperes)	MINIMUM SIZE	
	Copper wire No. (AWG)	Aluminum or copper-clad aluminum wire No. (AWG)
15	14	12
20	12	10
60	10	8
100	8	6
200	6	4
300	4	2
400	3	1

© 2018 IRC®, International Code Council®

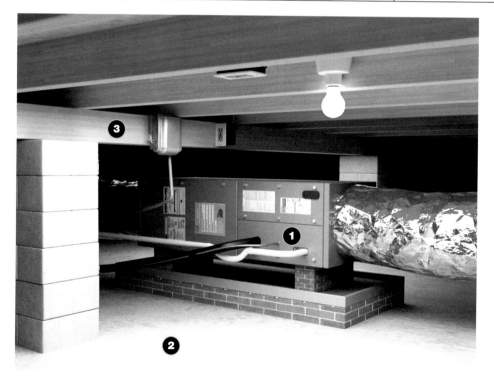

ELECTRICAL CONSIDERATIONS *(cont.)*

No.	Code	Description
	E3705.5.4	Air-conditioning and heat pump equipment circuit conductors shall be permitted to be protected against overcurrent in accordance with Section E3702.11 in the IRC code book.
❶	E3902.4, E3902.3, and E3902.5	Crawl space lighting and outlets for servicing HVAC equipment outside, in unfinished basements, and in crawl spaces must be GFCI type. All 125-volt, single-phase, 15- and 20-ampere receptacles installed outdoors shall have ground-fault circuit-interrupter protection for personnel. Where a crawl space is at or below grade level, all 125-volt, single-phase, 15- and 20-ampere receptacles installed in such spaces shall have ground-fault circuit-interrupter protection for personnel. All 125-volt, single-phase, 15- and 20-ampere receptacles installed in unfinished basements shall have ground-fault circuit-interrupter protection for personnel.
❷	E3903.4	In attics, under-floor spaces, utility rooms, and basements, at least one lighting outlet shall be installed where these spaces are used for storage or contain equipment requiring servicing. Such lighting outlet shall be controlled by a wall switch or shall have an integral switch. At least one point of control shall be at the usual point of entry to these spaces. The lighting outlet shall be provided at or near the equipment requiring servicing.
	E3904.7	If you use a stud cavity for air handling, the wire type and installation are regulated. Where wiring methods having a nonmetallic covering pass through stud cavities and joist spaces used for air handling, such wiring shall pass through such spaces perpendicular to the long dimension of the spaces.
❸	E3609.7 and G2411.1	Each aboveground portion of a gas piping system *other than* CSST that is likely to become energized shall be electrically continuous and bonded to an effective ground-fault current path. Where installed in or attached to a building or structure, metal piping systems, *including gas piping,* capable of becoming energized shall be bonded to the service equipment enclosure, the grounded conductor at the service, the grounding electrode conductor where of sufficient size, or to the one or more grounding electrodes used.

HVAC

CONDENSATE LINES

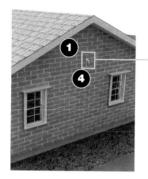
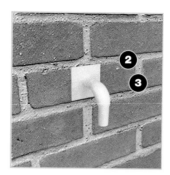
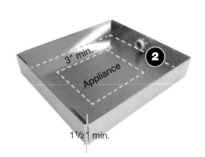

No.	Code	Description
1	M1411.3	Condensate from all cooling coils or evaporators shall be conveyed from the drain pan outlet to an approved place of disposal. Such piping shall maintain a minimum horizontal slope in the direction of discharge of not less than ⅛ unit vertical in 12 units horizontal (1% slope). Condensate shall not discharge into a street, alley, or other areas where it would cause a nuisance.
2	M1411.3.1	A secondary drain or auxiliary drain pan shall be required for each cooling or evaporator coil where damage to any building components will occur as a result of overflow from the equipment drain pan or stoppage in the condensate drain piping. Such piping shall maintain a minimum horizontal slope in the direction of discharge of not less than ⅛ unit vertical in 12 units horizontal.
3		Drain piping shall be a minimum of ¾″ nominal pipe size.
	M1411.3.1.1	On down-flow units and all other coils that have no secondary drain or provisions to install a secondary or auxiliary drain pan, a water-level monitoring device shall be installed inside the primary drain pan. This device shall shut off the equipment served in the event that the primary drain becomes restricted.
4	M1411.3.2	Components of the condensate disposal system shall be cast iron, galvanized steel, copper, polybutylene, polyethylene, ABS, CPVC, or PVC pipe or tubing.

You Should Know

- From the IRC code book, M1411.3.4: Where *appliances, equipment* or insulation are subject to water damage when auxiliary drain pans fill, those portions of the *appliances, equipment,* and insulation shall be installed above the flood level rim of the pan. Supports located inside of the pan to support the *appliance or equipment* shall be water-resistant and *approved*.

GAS APPLIANCE CONNECTIONS

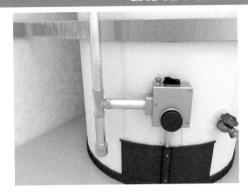

HVAC

Code	Description
G2419.1	Piping for other than dry gas conditions shall be sloped not less than 0.25″ in 15′ to prevent traps.
G2419.2	Where wet gas exists, a drip shall be provided at any point in the line of pipe where condensate could collect. A drip shall also be provided at the outlet of the meter and shall be installed so as to constitute a trap wherein an accumulation of condensate will shut off the flow of gas before the condensate will run back into the meter.
G2419.4	Sediment trap. Where a trap is not incorporated as part of the appliance, a sediment trap shall be installed downstream of the appliance shutoff valve as close to the inlet of the appliance as practical. The sediment trap shall be either a tee fitting having a capped nipple of any length installed vertically in the bottom-most opening of the tee, as illustrated in Figure G2419.4 in the IRC code book, or other device approved as an effective sediment trap. Illuminating appliances, ranges, clothes dryers, decorative vented appliances for installation in vented fireplaces, gas fireplaces, and outdoor grills need not be so equipped.
G2420.2	Every meter shall be equipped with a shutoff valve located on the supply side of the meter.
G2420.5.1	The shutoff valve shall be within 6′ of the appliance, and shall be installed upstream of the union, connector, or quick disconnect device it serves. Such shutoff valves shall be provided with access. Appliance shutoff valves located in the firebox of a fireplace shall be installed in accordance with the appliance manufacturer's instructions.
G2422.1.2.1	Connectors shall not exceed 6′ (1829 mm) in overall length. Measurement shall be made along the centerline of the connector. Only one connector shall be used for each appliance.

UNDERGROUND OIL STORAGE TANKS

No.	Code	Description
❶	M2201.3	Excavations for underground tanks shall not undermine the foundations of existing structures. The clearance from the tank to the nearest wall of a basement, pit, or property line shall not be less than 1'.
❷		Tanks shall be set on and surrounded with noncorrosive inert materials such as clean earth, sand, or gravel well tamped in place.
❸		Tanks shall be covered with not less than 1' of earth.
❹	M2203.4	Vent piping shall be not smaller than 1¼" pipe. Vent piping shall be laid to drain toward the tank without sags or traps in which the liquid can collect. Vent pipes shall not be cross-connected with fill pipes, lines from burners, or overflow lines from auxiliary tanks. The lower end of a vent pipe shall enter the tank through the top and shall extend into the tank not more than 1".
❺	M2203.5	Vent piping shall terminate outside of buildings at a point not less than 2', measured vertically or horizontally, from any building opening.

SECTION 3

PLUMBING

PLUMBING CONNECTIONS AND INSPECTIONS	
Code	**Description**
P2502.1	Where the entire sanitary drain system of an existing building is replaced, existing building drains under concrete slabs and existing building sewers that will serve the new system shall be internally examined to verify the piping is sloping in the correct direction, is not broken, is not obstructed, and is sized for the drainage load of the new plumbing drainage system to be installed.
P2502.2	Additions, alterations, renovations, or repairs to any plumbing system shall conform to that required for a new plumbing system without requiring the existing plumbing system to comply with all the requirements of this code. Additions, alterations, or repairs shall not cause an existing system to become unsafe, unsanitary, or overloaded.
P2503.2	A plumbing or drainage system, or part thereof, shall not be covered, concealed, or put into use until it has been tested, inspected, and approved by the building official.
P2503.4	The building sewer shall be tested by insertion of a test plug at the point of connection with the public sewer and pressurizing the sewer, filling the building sewer with water, testing with not less than a 10′ head of water, and be able to maintain such pressure for 15 minutes.
P2503.5.1	Drain, waste, and vent (DWV) systems shall be tested on completion of the rough piping installation by water or air, except air cannot be used for plastic piping, with no evidence of leakage. Either test shall be applied to the drainage system in its entirety or in sections after rough piping has been installed, as follows: 1. Water Test: Each section shall be filled with water to a point not less than 5′ above the highest fitting connection in that section, or to the highest point in the completed system. Water shall be held in the section under test for a period of 15 minutes. The system shall prove leak-free by visual inspection. 2. Air Test: The portion under test shall be maintained at a gauge pressure of 5 pounds per square inch (psi) or 10″ of mercury column. This pressure shall be held without introduction of additional air for a period of 15 minutes.

P2503.5.2	After the plumbing fixtures have been set and their traps filled with water, their connections shall be tested and proved gas tight and/or water tight as follows:
	1. Water Tightness: Each fixture shall be filled and then drained. Traps and fixture connections shall be proven water tight by visual inspection.
	2. Gas Tightness: When required by the local administrative authority, a final test for gas tightness of the DWV system shall be made by the smoke or peppermint test as follows:
	  1. Smoke Test: Introduce a pungent, thick smoke into the system. When the smoke appears at vent terminals, such terminals shall be sealed and a pressure equivalent to a 1″ water column shall be applied and maintained for a test period of not less than 15 minutes.
	  2. Peppermint Test: Introduce 2 ounces of oil of peppermint into the system. Add 10 quarts of hot water and seal all vent terminals. The odor of peppermint shall not be detected at any trap or other point in the system.

PLUMBING CONNECTIONS AND INSPECTIONS *(cont.)*

Code	Description
P2503.6	Where shower floors and receptors are made water tight by the application of materials required by Section P2709.2, the completed liner installation shall be tested. The pipe from the shower drain shall be plugged water tight for the test. The floor and receptor area shall be filled with potable water to a depth of not less than 2″ measured at the threshold. Where a threshold of at least 2″ high does not exist, a temporary threshold shall be constructed to retain the test water in the lined floor or receptor area to a level not less than 2″ deep measured at the threshold. The water shall be retained for a test period of not less than 15 minutes and there shall be no evidence of leakage.
P2503.7	Upon completion of the water-supply system or a section of it, the system or portion completed shall be tested and proved tight under a water pressure of not less than the working pressure of the system or, for piping systems other than plastic, by an air test of not less than 50 psi. This pressure shall be held for not less than 15 minutes. The water used for tests shall be obtained from a potable water source. See exception for PEX tested with compressed gas as authorized by manufacturer.
P2503.8	Inspection and testing of backflow prevention devices shall comply with Sections P2503.8.1 and P2503.8.2. P2503.8.1 Inspections: Inspections shall be made of all backflow prevention assemblies to determine whether they are operable. P2503.8.2 Testing: Reduced-pressure principle backflow preventers, double check valve assemblies, double-detector check valve assemblies, and pressure vacuum breaker assemblies shall be tested at the time of installation, immediately after repairs or relocation and at least annually.
P2503.9	Gauges used for testing shall be as follows: 1. Tests requiring a pressure of 10 psi or less shall utilize a testing gauge having increments of 0.10 psi or less. 2. Tests requiring a pressure higher than 10 psi but less than or equal to 100 psi shall use a testing gauge having increments of 1 psi or less. 3. Tests requiring a pressure higher than 100 psi shall use a testing gauge having increments of 2 psi or less.

PLUMBING

PIPE FITTINGS

Cap is used to close off the flow in a pipe.

Plug is used to close the flow in a pipe. It could have a threaded connection to allow it to serve as a cleanout or other fitting.

Coupling is used to join two lengths of pipe together.

Union is also used to join two lengths of pipe together.

90° elbow is used to turn the flow in a pipe 90° by turning the angle in the pipe. This fitting is also known as a *quarter bend*.

Street ell is used to join a piece of pipe and another fitting at an angle. It has a female fitting on one end and a male fitting or pipe thread on the other.

Reducing elbow is used, as the name implies, to reduce one end smaller than the other end. This fitting is used to connect two different pipe sizes while adjusting the flow 90°.

45° elbow is used to adjust the flow 45°. This is also called an *eighth bend*.

Street 45° elbow is used to turn the pipe or fitting 45°. In a street elbow one end is female and the other end is male.

PIPE FITTINGS *(cont.)*

 22½° elbow is used to turn the pipe 22½°. This fitting is also called a 1/16th bend.

 Tee is used to connect three separate lengths of pipe. The tee could also be used to establish a branch off of a continuous line.

 Reducing tee is used to connect up to three different-sized pipes together. The reducing tee has a nomenclature based on the size of each end—flow end large-size first followed by the other end and the branch last. If the two ends are the same, there is no need to repeat the second size.

 Sanitary tee is used for drain, waste, and vent only. It cannot be installed on its back or horizontally or an unsanitary condition could be created.

 Combination Y and eighth bend is a fitting used for drain, waste, and vent only. It is used as transition flow from one direction to a right angle in a gentle manner.

 Male adapter is a coupling that has male threads on one end with the opposite end adapting to the type of pipe used.

 Female adapter is a coupling that has female threads on one end with the opposite end adapting to the type of pipe used.

 P trap is a trap used to form a water seal to prevent sewer gases from entering the home.

 Closet flange is used to transition a water closet (toilet) to the closet band (another fitting) in the drainage system.

PLUMBING

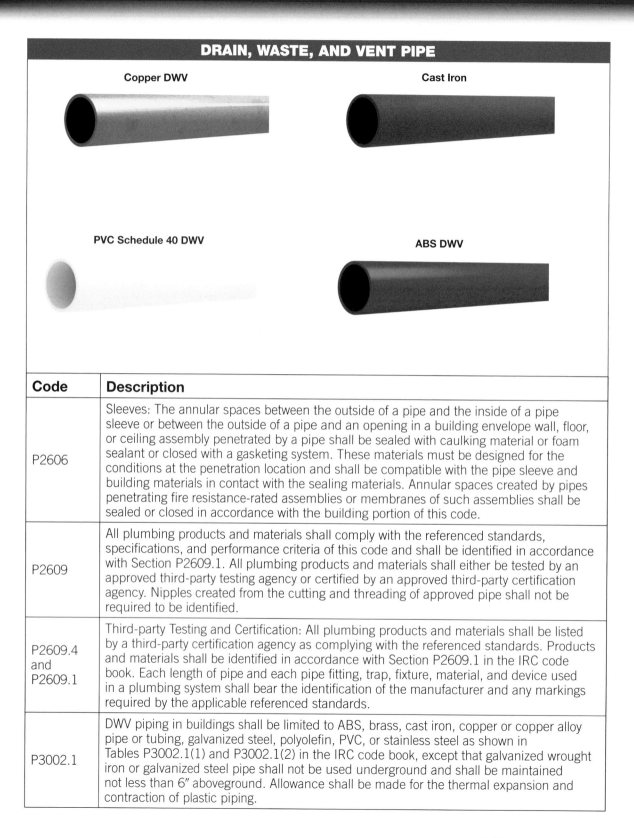

DRAIN, WASTE, AND VENT PIPE

Copper DWV

Cast Iron

PVC Schedule 40 DWV

ABS DWV

Code	Description
P2606	Sleeves: The annular spaces between the outside of a pipe and the inside of a pipe sleeve or between the outside of a pipe and an opening in a building envelope wall, floor, or ceiling assembly penetrated by a pipe shall be sealed with caulking material or foam sealant or closed with a gasketing system. These materials must be designed for the conditions at the penetration location and shall be compatible with the pipe sleeve and building materials in contact with the sealing materials. Annular spaces created by pipes penetrating fire resistance-rated assemblies or membranes of such assemblies shall be sealed or closed in accordance with the building portion of this code.
P2609	All plumbing products and materials shall comply with the referenced standards, specifications, and performance criteria of this code and shall be identified in accordance with Section P2609.1. All plumbing products and materials shall either be tested by an approved third-party testing agency or certified by an approved third-party certification agency. Nipples created from the cutting and threading of approved pipe shall not be required to be identified.
P2609.4 and P2609.1	Third-party Testing and Certification: All plumbing products and materials shall be listed by a third-party certification agency as complying with the referenced standards. Products and materials shall be identified in accordance with Section P2609.1 in the IRC code book. Each length of pipe and each pipe fitting, trap, fixture, material, and device used in a plumbing system shall bear the identification of the manufacturer and any markings required by the applicable referenced standards.
P3002.1	DWV piping in buildings shall be limited to ABS, brass, cast iron, copper or copper alloy pipe or tubing, galvanized steel, polyolefin, PVC, or stainless steel as shown in Tables P3002.1(1) and P3002.1(2) in the IRC code book, except that galvanized wrought iron or galvanized steel pipe shall not be used underground and shall be maintained not less than 6" aboveground. Allowance shall be made for the thermal expansion and contraction of plastic piping.

DRAIN, WASTE, AND VENT PIPE *(cont.)*	
Code	**Description**
Tables P3002.1(1) and (2)	Cast-iron pipe is approved for both aboveground and belowground installations.
	Copper or copper-alloy pipe is approved for aboveground installations and copper or copper-alloy tubing (Type K, L, M, or DWV) is approved for aboveground and belowground installations.
	Polyvinyl chloride (PVC) plastic pipe in IPS diameters, including schedule 40, DR 22 (PS 200), and DR 24 (PS 140) with a solid, cellular core or composite wall ASTM D 2665 and polyvinyl chloride (PVC) plastic pipe with a 3.25" O.D. and a solid, cellular core or composite wall is approved for both aboveground and belowground installations.
	Acrylonitrile butadiene styrene (ABS) is approved for both aboveground or underground installations.

You Should Know

- All plumbing products and materials shall comply with the referenced standards, specifications, and performance criteria of this code and shall be identified in accordance with Chapter 26 in the IRC code book. Plumbing products and materials shall either be tested by an approved third-party testing agency or certified by an approved third-party certification agency.

WATER PIPE

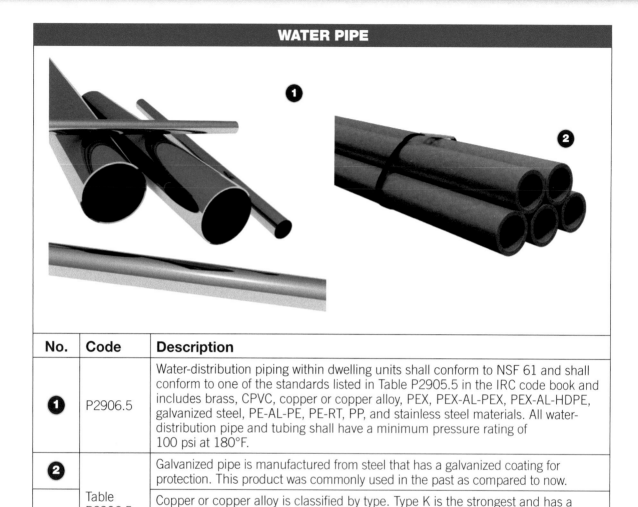

No.	Code	Description
❶	P2906.5	Water-distribution piping within dwelling units shall conform to NSF 61 and shall conform to one of the standards listed in Table P2905.5 in the IRC code book and includes brass, CPVC, copper or copper alloy, PEX, PEX-AL-PEX, PEX-AL-HDPE, galvanized steel, PE-AL-PE, PE-RT, PP, and stainless steel materials. All water-distribution pipe and tubing shall have a minimum pressure rating of 100 psi at 180°F.
❷	Table P2906.5	Galvanized pipe is manufactured from steel that has a galvanized coating for protection. This product was commonly used in the past as compared to now.
		Copper or copper alloy is classified by type. Type K is the strongest and has a green stripe for identification. Type L is less strong and has a blue stripe and can be used both aboveground and belowground. Type M is limited to aboveground uses in a building. It has a red stripe and is the lightest weight of the copper alloys.
		PVC schedule 40 is widely used as a water-service pipe. Generally, PVC is limited to cold water only.

You Should Know

- The type of cement for joining different types of plastic pipe is unique to the type of plastic. Be sure to use the correct cement and primer for sealing joints.

WATER PIPE

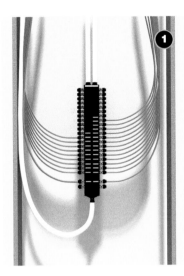

No.	Code	Description
❶	Table P2906.5	Cross-linked polyethylene (PEX), cross-linked polyethylene/aluminum/high-density polyethylene (PEX-AL-HDPE), and polyethylene/aluminum/polyethylene (PE-AL-PE) composite pipes are approved for use as water-distribution pipes. PEX is available in certain colors. The most common is red and blue (for hot and cold). It also comes in white and gray. The pipe is available in rolls and straight pipe sections. Connector fittings usually consist of an insert and a crimp ring with a special crimping device.
❷		Polypropylene (PP) plastic pipe or tubing is a newly accepted material allowed to carry potable water.
❸		Chlorinated polyvinyl chloride (CPVC) plastic pipe and tubing is cream in color to be distinct from PVC. CPVC pipe is approved for hot or cold water.

You Should Know

- There are other less commonly used pipe materials that are accepted for use with water distribution, including stainless-steel and brass pipe.

PLUMBING

PROTECTION AGAINST DAMAGE

0.0575" (1.463 mm)
thick metal nail plate

①

1¼"
(32 mm)
or less
from edge

No.	Code	Description
①	P2603.2.1	In concealed locations, where piping other than cast-iron or galvanized steel is installed through holes or notches in studs, joists, rafters, or similar members less than 1¼" from the nearest edge of the member, the pipe shall be protected by steel shield plates. Such shield plates shall have a thickness of not less than 0.0575" (16 gauge). Such plates shall cover the area of the pipe where the member is notched or bored and shall extend a minimum of 2" above sole plates and below top plates.
	P2603.3	Metallic piping except for cast iron, ductile iron, and galvanized steel shall not be placed in direct contact with steel framing members, concrete, or masonry. Metallic piping shall not be placed in direct contact with corrosive soil. Where sheathing is used to prevent direct contact, the sheathing material shall be not less than 0.008" or 8 mil and shall be made of plastic.
	P2603.4	Any pipe that passes through a concrete or masonry foundation wall shall be provided with a relieving arch or pipe sleeve built into the wall, the pipe sleeve being two pipe sizes greater than the pipe passing through.
	P2603.5	In localities having a winter design temperature of 32°F or lower, a water, soil, or waste pipe shall not be installed outside of a building, in exterior walls, in attics or crawl spaces, or in any other places subjected to freezing temperature unless adequate provision is made to protect it from freezing by insulation or heat or both. Water-service pipe shall be installed not less than 12" deep and not less than 6" below the frost line.

You Should Know

- Valves, pipes, and fittings shall be installed in correct relationship to the direction of the flow. Burred ends shall be reamed to the full bore of the pipe.

TRENCHING AND BACKFILLING

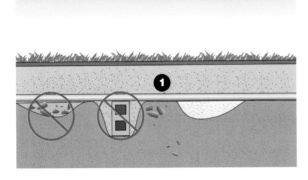

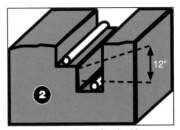

Trench with shelf

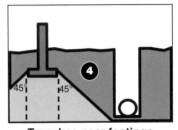

Pipes in same trench

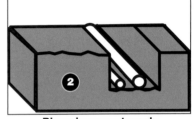

Trenches near footings

Figure P2604.4
Pipe Location with Respect to Footings

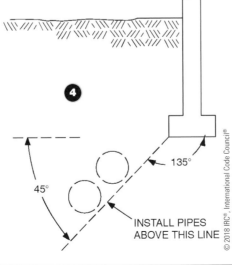

135°

45°

INSTALL PIPES
ABOVE THIS LINE

© 2018 IRC®, International Code Council®

PLUMBING

TRENCHING AND BACKFILLING *(cont.)*

No.	Code	Description
1	P2604.1	Where trenches are excavated such that the bottom of the trench forms the bed for the pipe, solid and continuous load-bearing support shall be provided between joints. Where overexcavated, the trench shall be backfilled to the proper grade with compacted earth, sand, fine gravel, or similar granular material. Piping shall not be supported on rocks or blocks at any point. Rocky or unstable soil shall be overexcavated by two or more pipe diameters and brought to the proper grade with suitable compacted granular material.
2	P2906.4.1	Trenching, pipe installation, and backfilling shall be in accordance with Section P2604 in the IRC code book. Water-service pipe is permitted to be located in the same trench with a building sewer, provided such sewer is constructed of materials listed for underground use within a building in Section P3002.1 in the IRC code book. If the building sewer is not constructed of materials listed in Section P3002.1 in the IRC code book, the water-service pipe shall be separated from the building sewer by a minimum of 5', measured horizontally, of undisturbed or compacted earth or placed on a solid ledge at least 12" above and to one side of the highest point in the sewer line.
3	P2604.3	Backfill shall be free from discarded construction material and debris. Backfill shall be free from rocks, broken concrete, and frozen chunks until the pipe is covered by at least 12" of tamped earth. Backfill shall be placed evenly on both sides of the pipe and tamped to retain proper alignment. Loose earth shall be carefully placed in the trench in 6" layers and tamped in place.
4	P2604.4	Trenching installed parallel to footings shall not extend into the bearing plane of footing or wall. The upper boundary of the bearing plane is a line that extends downward at an angle of 45 degrees from horizontal, from the outside, bottom edge of the footing or wall.

SUPPORT

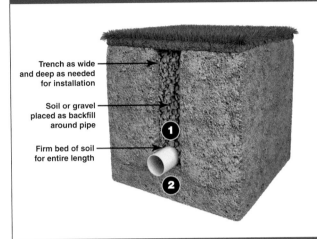

Trench as wide and deep as needed for installation

Soil or gravel placed as backfill around pipe

Firm bed of soil for entire length

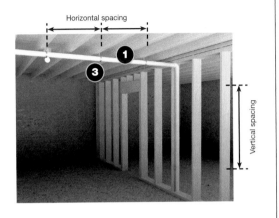

Horizontal spacing

Vertical spacing

TABLE P2605.1 PIPING SUPPORT

❹ PIPING MATERIAL	MAXIMUM HORIZONTAL SPACING (feet)	MAXIMUM VERTICAL SPACING (feet)
ABS pipe	4	10[b]
Aluminum tubing	10	15
Cast-iron pipe	5[a]	15
Copper or copper alloy pipe	12	10
Copper or copper alloy tubing (1¼ inches in diameter and smaller)	6	10
Copper or copper alloy tubing (1½ inches in diameter and larger)	10	10
Cross-linked polyethylene (PEX) pipe, 1 inch and smaller	2.67 (32 inches)	10[b]
Cross-linked polyethylene (PEX) pipe, 1¼ inches and larger	4	10[b]
Cross-linked polyethylene/aluminum/cross-linked polyethylene (PEX-AL-PEX) pipe	2.67 (32 inches)	4[b]
CPVC pipe or tubing (1 inch in diameter and smaller)	3	10[b]
CPVC pipe or tubing (1¼ inches in diameter and larger)	4	10[b]
Lead pipe	Continuous	4
PB pipe or tubing	2.67 (32 inches)	4
Polyethylene of raised temperature (PE-RT) pipe, 1 inch and smaller	2.67 (32 inches)	10[b]
Polyethylene of raised temperature (PE-RT) pipe, 1¼ inch and larger	4	10[b]
Polypropylene (PP) pipe or tubing (1 inch and smaller)	2.67 (32 inches)	10[b]
Polypropylene (PP) pipe or tubing (1¼ inches and larger)	4	10[b]
PVC pipe	4	10[b]
Stainless steel drainage systems	10	10[b]
Steel pipe	12	15

For SI: 1 inch = 25.4 mm, 1 foot = 304.8 mm.

a. The maximum horizontal spacing of cast-iron pipe hangers shall be increased to 10 feet where 10-foot lengths of pipe are installed.
b. For sizes 2 inches and smaller, a guide shall be installed midway between required vertical supports. Such guides shall prevent pipe movement in a direction perpendicular to the axis of the pipe.

© 2018 IRC®, International Code Council®

PLUMBING

No.	Code	Description
		SUPPORT *(cont.)*
		Piping shall be supported in accordance with the following:
❶		Piping shall be supported to ensure alignment and prevent sagging and allow movement associated with the expansion and contraction of the piping system.
❷		Piping in the ground shall be laid on a firm bed for its entire length, except where support is otherwise provided.
❸	P2605.1	Hangers and anchors shall be of sufficient strength to maintain their proportional share of the weight of pipe and contents and of sufficient width to prevent distortion to the pipe. Hangers and strapping shall be of approved material that will not promote galvanic action. Rigid support sway bracing shall be provided at changes in direction greater than 45° for pipe sizes 4″ and larger.
		Where horizontal pipes 4″ and larger convey drainage or waste and where a pipe fitting changes flow direction greater than 45°, rigid bracing or similar support shall be installed to resist movement of the upstream pipe in the direction of the flow.
❹		Piping shall be supported at distances not to exceed those indicated in Table P2605.1.

DRAIN, WASTE, AND VENT PIPE

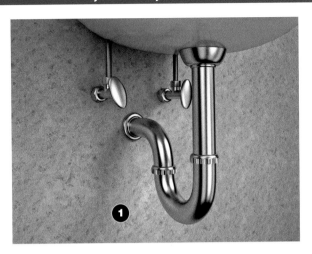

No.	Code	Description
❶	P2703	Fixture tail pieces shall be not less than 1½″ (38 mm) in diameter for sinks, dishwashers, laundry tubs, bathtubs, and similar fixtures, and not less than 1¼″ in diameter for bidets, lavatories, and similar fixtures.
❷	P2704	Slip joint connections shall be installed only for tubular waste piping and only between the trap outlet and the fixture and the connection to the drainage piping. Slip joint shall be accessible. Such access shall be 12″ in the least direction.

PLUMBING

FIXTURE INSTALLATION

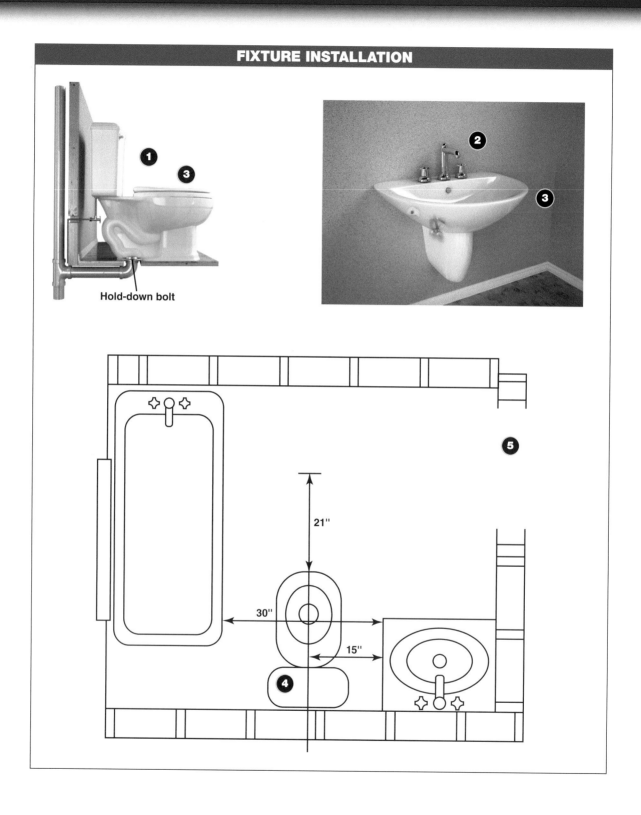

Hold-down bolt

21"

30"

15"

FIXTURE INSTALLATION *(cont.)*

No.	Code	Description
❶		Floor-outlet or floor-mounted fixtures shall be secured to the drainage connection and to the floor, where so designed, by screws, bolts, washers, nuts, and similar fasteners of copper alloy or other corrosion-resistant material.
❷		Wall-hung fixtures shall be rigidly supported so that strain is not transmitted to the plumbing system.
❸		Where fixtures come in contact with walls and floors, the contact area shall be water tight.
		Plumbing fixtures shall be usable.
❹	P2705.1 Items 1–8	Water Closets, Lavatories, and Bidets: A water closet, lavatory, or bidet shall not be set closer than 15" from its center to any side wall, partition, or vanity or closer than 30" center-to-center between adjacent fixtures. There shall be at least a 21" clearance in front of the water closet, lavatory, or bidet to any wall, fixture, or door.
❺		The location of piping, fixtures, or equipment shall not interfere with the operation of windows or doors.
		In areas prone to flooding, plumbing fixtures shall be located or installed in accordance with Section R322.1.7 in the IRC code book.
		Integral fixture-fitting mounting surfaces on manufactured plumbing fixtures or plumbing fixtures constructed on site shall meet the design requirements of ASME A112.19.2 or ASME A112.19.3.

PLUMBING

WASTE RECEPTORS

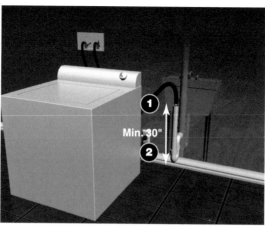

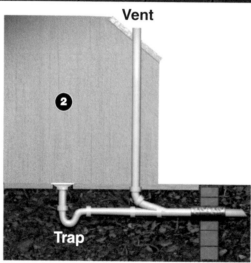

No.	Code	Description
❶		For other than hub drains that receive only clear-water waste and standpipes, a removable strainer or basket shall cover the waste outlet of waste receptors. Waste receptors shall not be installed in concealed spaces.
❷	P2706.1	Waste receptors shall be readily accessible.
		Waste receptors shall not be installed in plenums, attics, crawl spaces, interstitial spaces above ceilings, and below floors.
		Ready access shall be provided to waste receptors.

DIRECTIONAL FITTINGS

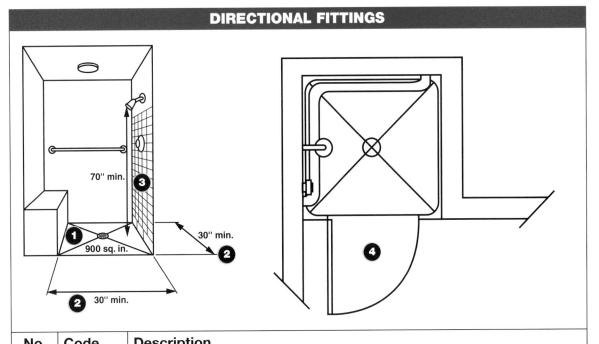

No.	Code	Description
❶		Shower compartments shall have at least 900 in² of interior cross-sectional area.
❷		Shower compartments shall be not less than 30″ in minimum dimension measured from the finished interior dimension of the shower compartment, exclusive of fixture valves, showerheads, soap dishes, and safety grab bars or rails.
❸	P2708.1	The minimum required area and dimension shall be measured from the finished interior dimension at a height equal to the top of the threshold and at a point tangent to its centerline and shall be continued to a height of not less than 70″ above the shower drain outlet.
❹		Hinged shower doors shall open outward.
		The wall area above built-in tubs having installed showerheads and in shower compartments shall be constructed in accordance with Section R702.4 in the IRC code book. Such walls shall form a water tight joint with each other and with either the tub, receptor, or shower floor.

You Should Know

- Fold-down seats shall be permitted in the shower, provided the required 900 in² dimension is maintained when the seat is in the folded-up position.
- Shower compartments having not less than 25″ in minimum dimension measured from the finished interior dimension of the compartment are permitted, provided that the shower compartment has a minimum of 1300 in² of cross-sectional area.

PLUMBING

SHOWER RECEPTORS

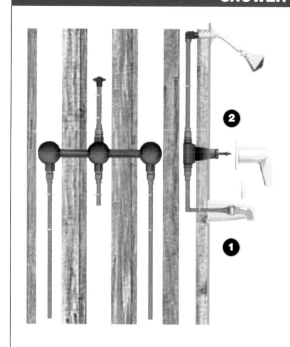

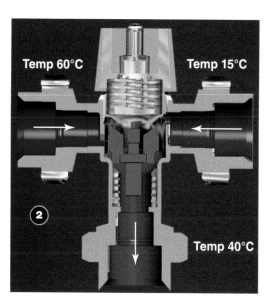

Temp 60°C

Temp 15°C

Temp 40°C

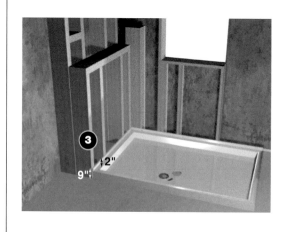

9" 2"

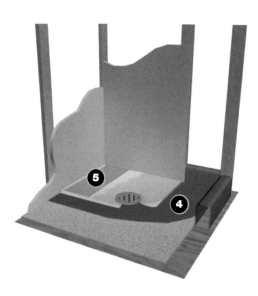

No.	Code	Description
1	P2708.3	Water-supply risers from the shower valve to the showerhead outlet, whether exposed or concealed, shall be attached to the structure using support devices designed for use with the specific piping material or fittings and anchored with screws.
2	P2708.4	Individual shower and tub/shower combination valves shall be equipped with control valves of the pressure-balance, thermostatic-mixing, or combination pressure-balance/thermostatic-mixing valve types with a high-limit stop that shall be set to limit water temperature to a maximum of 120°F.
3	P2709.1	Where a shower receptor has a finished curb threshold, it shall be not less than 1″ below the sides and back of the receptor. The curb shall be not less than 2″ and not more than 9″ deep when measured from the top of the curb to the top of the drain. The finished floor shall slope uniformly toward the drain not less than ¼ unit vertical in 12 units horizontal nor more than ½ unit vertical per 12 units horizontal (4% slope), and floor drains shall be flanged to provide a water tight joint in the floor.
4	P2709.2	The adjoining walls and floor framing enclosing on-site, built-up shower receptors shall be lined and extend not less than 3″ beyond or around the rough jambs and not less than 3″ above finished thresholds. Sheet-applied, load-bearing, bonded waterproof membranes shall be applied in accordance with the manufacturer's installation instructions.
5	P2709.3	Lining materials shall be pitched ¼ unit vertical in 12 units horizontal to weep holes in the subdrain by means of a smooth, solidly formed sub-base and shall be properly recessed and fastened to approved backing so as not to occupy the space required for the wall covering and shall not be nailed or perforated at any point less than 1″ above the finished threshold.

SHOWER RECEPTORS (cont.)

PLUMBING

LAVATORIES

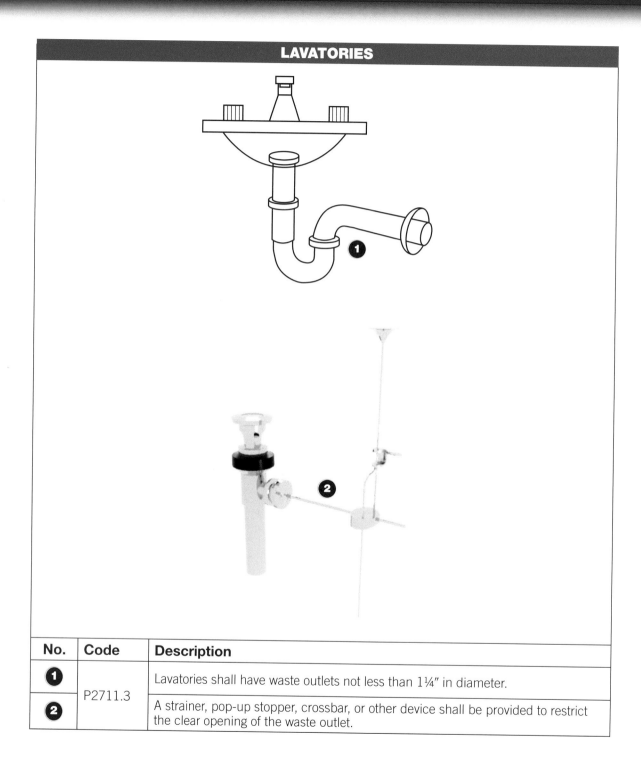

No.	Code	Description
1	P2711.3	Lavatories shall have waste outlets not less than 1¼" in diameter.
2		A strainer, pop-up stopper, crossbar, or other device shall be provided to restrict the clear opening of the waste outlet.

WATER CLOSETS

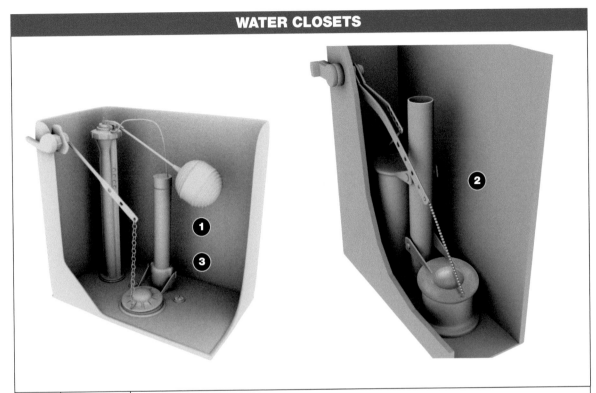

No.	Code	Description
①	P2712.2	Water closets shall be provided with a flush tank, flushometer tank, or flushometer valve designed and installed to supply water in sufficient quantity and flow to flush the contents of the fixture, to cleanse the fixture, and refill the fixture trap in accordance with ASME A112.19.2 and ASME A112.19.6 and other related standards.
②	P2712.4	Flush valve seats in flush tanks for flushing water closets shall be at least 1" above the flood-level rim of the bowl connected thereto, except an approved water closet and flush tank combination designed so that when the tank is flushed and the fixture is clogged or partially clogged, the flush valve will close tightly so that water will not spill continuously over the rim of the bowl or backflow from the bowl to the tank.
③	P2712.6	All parts in a flush tank shall be accessible for repair and replacement.
	P2712.7	Water closets shall be equipped with seats of smooth, nonabsorbent material and shall be properly sized for the water closet bowl type.

PLUMBING

BATHTUBS, SINKS, AND LAUNDRY TUBS

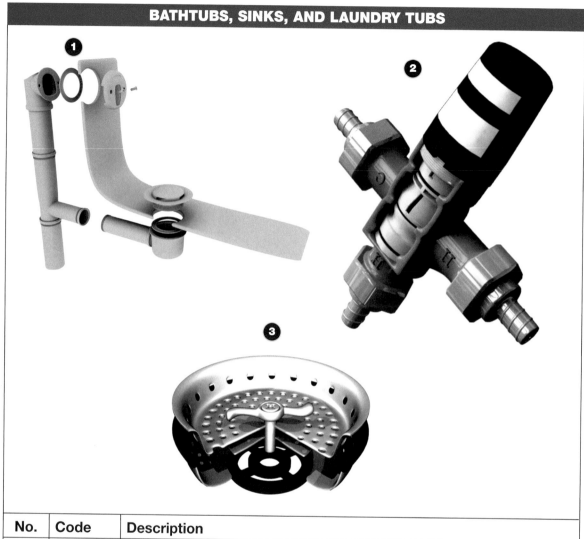

No.	Code	Description
❶	P2713.1	Bathtubs shall be equipped with a waste outlet. The outlets shall be connected to waste tubing or piping not less than 1½" in diameter. The waste outlet shall be equipped with a water tight stopper. Where an overflow is installed, the overflow shall not be less than 1½" in diameter.
❷	P2713.3	The hot water supplied to bathtubs and whirlpool bathtubs shall be limited to a maximum temperature of 120°F by a water-temperature-limiting device that conforms to ASSE 1070 or CSA B125.3, except where such protection is otherwise provided by a combination tub/shower valve in accordance with Section P2708.3 in the IRC code book.
❸	P2714.1	Sinks shall be provided with waste outlets not less than 1½" in diameter. A strainer, crossbar, or other device shall be provided to restrict the clear opening of the waste outlet.
	P2715.1	Each compartment of a laundry tub shall be provided with a waste outlet not less than 1½" in diameter and a strainer or crossbar to restrict the clear opening of the waste outlet.

FOOD-WASTE DISPOSER, DISHWASHERS, AND CLOTHES WASHERS

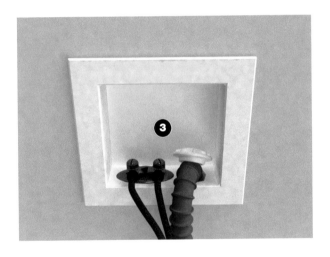

No.	Code	Description
1	P2716	Food-waste disposers shall be connected to a drain of not less than 1½" in diameter.
2	P2717.2	The combined discharge from a one- or two-compartment sink, dishwasher, and waste disposer is permitted to discharge through a single 1½" trap. The discharge pipe from the dishwasher shall rise to the underside of the counter and be fastened or otherwise held in that position before connecting to the head of the food waste disposer or to the wye fitting in the sink tailpiece.
3	P2718.1	The discharge from a clothes washing machine shall be through an air break.

PLUMBING

FLOOR DRAINS AND WHIRLPOOL TUBS

No.	Code	Description
1	P2719.1	Floor drains shall have waste outlets not less than 2" diameter and a removable strainer. The floor drain shall be constructed so that the drain can be cleaned. Access shall be provided to the drain inlet. Floor drains shall not be located under or have their access restricted by permanently installed appliances.
2	P2720.1	Access shall be provided to circulation pumps in accordance with the fixture or pump manufacturer's installation instructions. Where the manufacturer's instructions do not specify the location and minimum size of field-fabricated access openings, a 12" by 12" minimum size opening shall be installed for access to the circulation pump.
3		Where pumps are located more than 2' from the access opening, an 18" by 18" minimum size opening shall be installed.
4		A door or panel shall be permitted to close the opening. In all cases, the access opening shall be unobstructed and be of the size necessary to permit the removal and replacement of the circulation pump.

BIDETS AND FIXTURE FITTINGS

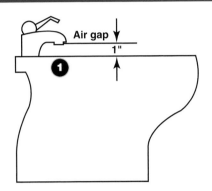

No.	Code	Description
1	P2721.1	The bidet shall be equipped with either an air-gap-type or vacuum-breaker-type fixture supply fitting.
2	P2721.2	The discharge water temperature from a bidet fitting shall be limited to a maximum temperature of 110°F by a water-temperature-limiting device conforming to ASSE 1070 CSA B125.3.
3	P2722.2	Fixture fittings and faucets that are supplied with both hot and cold water shall be installed and adjusted so that the left-hand side of the water-temperature control represents the flow of hot water when facing the outlet.
4	P2722.3	Faucets and fixture fittings with hose-connected outlets shall conform to ASME A112.18.3 or CSA B125.

PLUMBING

PAN

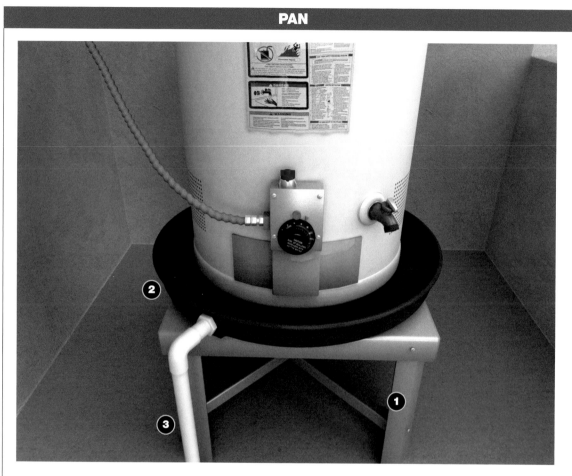

No.	Code	Description
1	P2801.6	Where a storage-tank-type water heater or a hot-water storage tank is installed in a location where water leakage from the tank will cause damage, the tank shall be installed in a galvanized steel pan having a material thickness of not less than 0.0236" (No. 24 gauge) or plastic not less than 0.036" in thickness or other approved materials. Plastic pans under gas water heaters must have a flame spread of 25 or less and a smoke developed index of no more than 450.
2	P2801.6.1	The pan shall be not less than 1½" deep and shall be of sufficient size and shape to receive all dripping or condensate from the tank or water heater. The pan shall be drained by an indirect waste pipe having a minimum diameter of ¾".
3	P2801.6.2	The pan drain shall extend full size and terminate over a suitably located indirect waste receptor or shall extend to the exterior of the building and terminate not less than 6" and not more than 24" above the adjacent ground surface.

WATER HEATER INSTALLATION

Garage gas water heater installation

Protective bollard

No.	Code	Description
❶	M1307.3.1	Protection from Impact: Appliances shall not be installed in a location subject to vehicle damage, except where protected by approved barriers.
❷	P2801.7	Water heaters having an ignition source shall be elevated such that the source of ignition is not less than 18″ above the garage floor, except those that are flammable vapor ignition-resistant.
❸	P2801.8	In Seismic Design Categories D_0, D_1, and D_2 and townhouses in Seismic Design Category C, water heaters shall be anchored or strapped in the upper one-third and in the lower one-third of the appliance to resist a horizontal force equal to one-third of the operating weight of the water heater, acting in any horizontal direction, or in accordance with the appliance manufacturer's recommendations.

PLUMBING

RELIEF VALVES

Discharge Pipe Section P2804.6.1

The discharge piping serving a pressure-relief valve, temperature-relief valve, or combination valve shall:

1. Not be directly connected to the drainage system
2. Discharge through an air gap located in the same room as the water heater
3. Not be smaller than the diameter of the outlet of the valve served and shall discharge full size to the air gap
4. Serve a single relief device and shall not connect to piping serving any other relief device or equipment
5. Discharge to the floor, to the pan serving the water heater or storage tank, to a waste receptor, or to the outdoors
6. Discharge in a manner that does not cause personal injury or structural damage
7. Discharge to a termination point that is readily observable by the building occupants
8. Not be trapped
9. Be installed to flow by gravity
10. Not terminate more than 6″ and not less than 2 times the discharge pipe diameter above the floor or waste receptor
11. Not have a threaded connection at the end of the piping
12. Not have valves or tee fittings
13. Be constructed of those materials listed in Section P2904.5 in the IRC code book or materials tested, rated, and *approved* for such use in accordance with ASME A112.4.1
14. Be one nominal size larger than the size of the relief-valve outlet, where the relief-valve discharge piping is constructed of PEX or PE-RT tubing

RELIEF VALVES *(cont.)*	
Code	**Description**
P2804.1	Appliances and equipment used for heating water or storing hot water shall be protected by: 1. A separate pressure-relief valve and a separate temperature-relief valve; or 2. A combination pressure- and temperature-relief valve.
P2804.4	Temperature-relief valves shall have a temperature-sensing element to monitor the water within the top 6″ of the tank. The valve shall be set to open at a maximum temperature of 210°F.
P2804.6	A check or shutoff valve shall *not* be installed in the following locations: 1. Between a relief valve and the termination point of the relief valve discharge pipe; 2. Between a relief valve and a tank; or 3. Between a relief valve and heating appliances or equipment.

PLUMBING

CROSS-CONNECTION: PROTECTION OF POTABLE WATER SUPPLY

No.	Code	Description
	P2902.3	A means of protection against backflow shall be provided in accordance with Sections P2902.3.1 through P2902.3.6 in the IRC code book. Backflow prevention applications must guard against backsiphonage and backpressure based on degree of hazard. The proper device is based on this degree of hazard.
	P2902.4.3	Sillcocks, hose bibbs, wall hydrants, and other openings with a hose connection shall be protected by an atmospheric-type or pressure-type vacuum breaker or a permanently attached hose connection vacuum breaker.
❶	P2902.5.3	The potable water supply to lawn irrigation systems shall be protected against backflow by an atmospheric vacuum breaker, a pressure vacuum breaker, or a reduced-pressure principle backflow preventer.
		A valve shall not be installed downstream from an atmospheric vacuum breaker.
		Where chemicals are introduced into the system, the potable water supply shall be protected against backflow by a reduced-pressure principle backflow preventer.

You Should Know

- Backflow of non-potable water into the potable water system is prevented with a variety of devices or means, including an air gap, atmospheric-type vacuum breaker, backflow preventer with intermediate atmospheric vent, pressure-type vacuum breakers, reduced-pressure principle backflow preventers, and double check-valve assemblies.

WATER SUPPLY AND DISTRIBUTION

TABLE P2903.2 MAXIMUM FLOW RATES AND CONSUMPTION FOR PLUMBING FIXTURES AND FIXTURE FITTINGS[b]

➊ PLUMBING FIXTURE OR FIXTURE FITTING	PLUMBING FIXTURE OR FIXTURE FITTING
Lavatory faucet	2.2 gpm at 60 psi
Shower head[a]	2.5 gpm at 80 psi
Sink faucet	2.2 gpm at 60 psi
Water closet	1.6 gallons per flushing cycle

For SI: 1 gallon per minute = 3.785 L/m, 1 pound per square inch = 6.895 kPa.

a. A handheld shower spray is also a shower head.
b. Consumption tolerances shall be determined from referenced standards.

© 2018 IRC®, International Code Council®

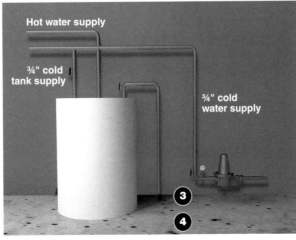

Hot water supply

¾" cold tank supply

¾" cold water supply

PLUMBING

No.	Code	Description
	P2903.1	The water-service and water-distribution systems shall be designed and pipe sizes shall be selected such that under conditions of peak demand, the capacities at the point of outlet discharge shall not be less than shown in Table P2903.1.
❶	P2903.2	The *maximum* water consumption flow rates and quantities for all plumbing fixtures and fixture fittings shall be in accordance with Table P2903.2.
❷	P2903.3	Where the water pressure supplied by the public water main or an individual water supply system is insufficient to provide minimum pressure and quantities for the plumbing fixtures in the building, the pressure shall be increased by means of an elevated water tank, hydropneumatic pressure booster system, or water pressure booster pump.
❸	P2903.3.1	Maximum static pressure shall be 80 psi. When main pressure exceeds 80 psi, an approved pressure-reducing valve conforming to ASSE 1003 shall be installed on the domestic water branch main or riser at the connection to the water-service pipe.
❹	P2903.4	A means for controlling increased pressure caused by thermal expansion shall be installed where required. For water-service system, sizes up to and including 2″ pressure-reducing device for controlling pressure shall be installed. Where a backflow prevention device, check valve, or other device is installed on a water-supply system using storage water-heating equipment such that thermal expansion causes an increase in pressure, a device for controlling pressure shall be installed.

WATER SUPPLY AND DISTRIBUTION *(cont.)*

WATER-SUPPLY PIPE SIZING

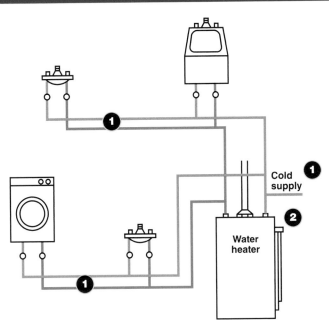

TABLE P2903.6 WATER-SUPPLY FIXTURE-UNIT VALUES FOR VARIOUS PLUMBING FIXTURES AND FIXTURE GROUPS

TYPE OF FIXTURES OR GROUP OF FIXTURES	WATER-SUPPLY FIXTURE-UNIT (w.s.f.u.) VALUE		
	Hot	Cold	Combined
Bathtub (with/without overhead shower head)	1.0	1.0	1.4
Clothes washer	1.0	1.0	1.4
Dishwasher	1.4	—	1.4
Full-bath group with bathtub (with/without shower head) or shower stall	1.5	2.7	3.6
Half-bath group (water closet and lavatory)	0.5	2.5	2.6
Hose bibb (sillcock)[a]	—	2.5	2.5
Kitchen group (dishwasher and sink with/without food-waste disposal)	1.9	1.0	2.5
Kitchen sink	1.0	1.0	1.4
Laundry group (clothes washer standpipe and laundry tub)	1.8	1.8	2.5
Laundry tub	1.0	1.0	1.4
Lavatory	0.5	0.5	0.7
Shower stall	1.0	1.0	1.4
Water closet (tank type)	—	2.2	2.2

For SI: 1 gallon per minute = 3.785 L/m.

a. The fixture unit value 2.5 assumes a flow demand of 2.5 gpm, such as for an individual lawn sprinkler device. If a hose bibb/sill cock will be required to furnish a greater flow, the equivalent fixture-unit value may be obtained from this table or Table P2903.6(1).

© 2018 IRC®, International Code Council®

PLUMBING

No.	Code	Description
1	P2903.6	Supply loads in the building water-distribution system shall be determined by total load on the pipe being sized in terms of water-supply fixture units (w.s.f.u.), as shown in Table P2903.6, and gallon per minute (gpm) flow rates.
2	P2903.7	The minimum-size water-service pipe shall be ¾″.
		The size of water-service mains, branch mains, and risers shall be determined according to water-supply demand, available water pressure, and friction loss caused by the water meter and developed length of pipe, including equivalent length of fittings.

WATER-SUPPLY PIPE SIZING (cont.)

You Should Know

- For fixtures not listed, choose a w.s.f.u. value of a fixture with similar flow characteristics.
- The size of each water-distribution system shall be determined according to design methods conforming to acceptable engineering practice, such as those methods in Appendix P in the IRC code book, and shall be approved by the code official.

WATER-SUPPLY VALVES

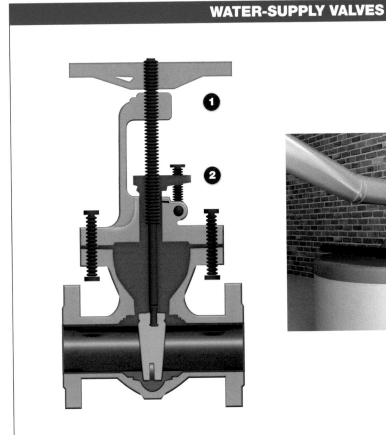

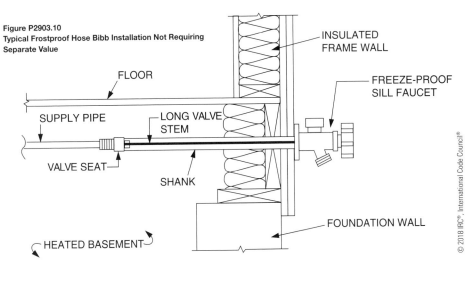

Figure P2903.10
Typical Frostproof Hose Bibb Installation Not Requiring
Separate Value

FLOOR

SUPPLY PIPE

LONG VALVE
STEM

VALVE SEAT

SHANK

HEATED BASEMENT

INSULATED
FRAME WALL

FREEZE-PROOF
SILL FAUCET

FOUNDATION WALL

© 2018 IRC®, International Code Council®

PLUMBING

No.	Code	Description
WATER-SUPPLY VALVES *(cont.)*		
❶	P2903.9.1	Each dwelling unit shall be provided with an accessible main shutoff valve near the entrance of the water service. The valve shall be of a full-open type having nominal restriction to flow, with provision for drainage such as a bleed orifice or installation of a separate drain valve. Additionally, the water service shall be valved at the curb or lot line in accordance with local requirements.
❷	P2903.9.2	A readily accessible full-open valve shall be installed in the cold-water supply pipe to each water heater at or near the water heater.
	P2903.9.3	Shutoff valves shall be required on each fixture supply pipe to each plumbing appliance and to each plumbing fixture other than bathtubs and showers. Valves serving individual fixtures, plumbing appliances, risers, and branches shall be provided access.
	P2903.9.5	Potable water outlets and combination stop-and-waste valves shall not be installed underground or below grade. Freeze-proof yard hydrants that drain the riser into the ground are considered to be stop-and-waste valves.
	P2903.10	Hose bibs subject to freezing, including the "frost-proof" type, shall be equipped with an accessible stop-and-waste-type valve inside the building so that they can be controlled and/or drained during cold periods.

You Should Know

- Frost-proof hose bibs installed such that the stem extends through the building insulation into an open-heated or semi-conditioned space need not be separately valved, see Figure P2903.10.

APPENDIX P—WATER PIPE SIZING

Code	Description
AP101.1	This appendix outlines two procedures for sizing a water piping system. The design procedures are based on the minimum static pressure available from the supply source, the head charges in the system caused by friction and elevation, and the rates of flow necessary for operation of various fixtures. The first method is more of a design method used by engineers. The second method uses tabular values for pipe size selection.
AP103.2.2 1st Method	Water pipe sizing procedures are based on a system of pressure requirements and losses, the sum of which must not exceed the minimum pressure available at the supply source. These pressures are as follows: 1. Pressure required at fixture to produce required flow. 2. Static pressure loss or gain (due to head) is computed at 0.433 psi per foot of elevation change. Example: Assume that the highest fixture supply outlet is 20' above or below the supply source. This produces a static pressure differential of 8.66 psi loss (20' by 0.433 psi/ft). 3. Loss through water meter. The friction or pressure loss can be obtained from meter manufacturers. 4. Loss through taps in water main. 5. Losses through special devices such as filters, softeners, backflow prevention devices, and pressure regulators. These values must be obtained from the manufacturers. 6. Loss through valves and fittings. Losses for these items are calculated by converting to *equivalent length* of piping and adding to the total pipe length. 7. Loss caused by pipe friction can be calculated when the pipe size, the pipe length, and the flow through the pipe are known. With these three items, the friction loss can be determined. For piping flow charts not included, use manufacturers' tables and velocity recommendations.
AP103.3	The size of water-service mains, branch mains, and risers must be determined by the segmented loss method according to water-supply demand (gpm), available water pressure (psi), and friction loss caused by the water meter and developed length of pipe (feet), including equivalent length of fittings. This design procedure is based on the following parameters: • The calculated friction loss through each length of the pipe • A system of pressure losses, the sum of which must not exceed the minimum pressure available at the street main or other source of supply • Pipe sizing based on estimated peak demand, total pressure losses caused by difference in elevation, equipment, developed length and pressure required at the most remote fixture, loss through taps in water main, losses through fittings, filters, backflow prevention devices, valves and pipe friction

PLUMBING

APPENDIX P—WATER PIPE SIZING *(cont.)*	
Code	**Description**

AP201 2nd Method	The minimum size water service pipe shall be ¾". The size of water service mains, branch mains, and risers shall be determined according to water-supply demand (gpm), available water pressure (psi), and friction loss caused by the water meter and developed length of pipe (feet), including equivalent length of fittings. The size of each water distribution system shall be determined according to the procedure outlined in this section or by other design methods conforming to acceptable engineering practice and approved by the code official:
	1. Supply load in the building water-distribution system shall be determined by total load on the pipe being sized, in terms of w.s.f.u., as shown in Table AP103.3(2) IRC in the IRC code book.
	2. Obtain the minimum daily static service pressure (psi) available (as determined by the local water authority) at the water meter or other source of supply at the installation location. Adjust this minimum daily static pressure (psi) for the following conditions:
	  1. Determine the difference in elevation between the source of supply and the highest water-supply outlet. Where the highest water-supply outlet is located above the source of supply, deduct 0.5 psi for each foot of difference in elevation. Where the highest water-supply outlet is located below the source of supply, add 0.5 psi for each foot of difference in elevation.
	  2. Where a water-pressure-reducing valve is installed in the water-distribution system, the minimum daily static water pressure available is 80% of the minimum daily static water pressure at the source of supply or the set pressure downstream of the pressure-reducing valve, whichever is smaller.
	  3. Deduct all pressure losses caused by special equipment such as a backflow preventer, water filter, and water softener. Pressure-loss data for each piece of equipment shall be obtained through the manufacturer of the device.
	  4. *Deduct the pressure in excess of 8 psi resulting from installation of the special plumbing fixture,* such as temperature-controlled shower and flushometer tank water closet. Using the resulting minimum available pressure, find the corresponding pressure range in Table AP201.1 found on page 248 of this text.
	3. The maximum developed length for water piping is the actual length of pipe between the source of supply and the most remote fixture, including either hot (through the water heater) or cold water branches multiplied by a factor of 1.2 to compensate for pressure loss through fittings.
	Select the appropriate column in Table AP201.1 equal to or greater than the calculated maximum developed length.
	4. To determine the size of water service pipe, meter, and main distribution pipe to the building using the appropriate table, follow down the selected "maximum developed length" column to a fixture unit equal to or greater than the total installation demand calculated by using the "combined" water-supply fixture unit column of Table AP201.1. Read the water service pipe and meter sizes in the first left-hand column and the main distribution pipe to the building in the second left-hand column on the same row.

APPENDIX P—WATER PIPE SIZING *(cont.)*

Code	Description
AP201 2nd Method	5. To determine the size of each water-distribution pipe, start at the most remote outlet on each branch (either hot or cold branch) and, working back toward the main distribution pipe to the building, add up the water-supply fixture unit demand passing through each segment of the distribution system using the related hot or cold column of Table AP201.1. Knowing demand, the size of each segment shall be read from the second left-hand column of the same table and maximum developed length column selected in Steps 1 and 2, under the same- or next-smaller-size meter row. In no case does the size of any branch or main need to be larger than the size of the main distribution pipe to the building established in Step 4.

You Should Know

- The most commonly used method is the second method that makes use of tables based on water-pressure ranges with deductions based on pressure losses.

PLUMBING

APPENDIX P—WATER PIPE SIZING *(cont.)*

Example

Given: A home has two full bathrooms (flush tank) and a half-bath (flush tank) with a kitchen sink and washing machine (8 lb load capacity). The maximum developed length of the water distribution system is 100′. The building is one story and the highest fixture outlet is 6′ above the source of the water supply. The minimum daily static pressure of the water supply is 72 psi at the meter. There are no pressure losses caused by special equipment and no pressure-reducing valves. There is one thermostatic mixing valve in the shower (required by code). The manufacturer reports that its pressure loss is 6.2 psi.

Step 1. Determine the water-supply fixture units (w.s.f.u.) from Table AP103.3(2).

Bathroom group: 3.6 (hot and cold) w.s.f.u. * 2 = 7.2
Kitchen sink: 1.4 (hot and cold) w.s.f.u. * 1 = 1.4
Water closet: 2.2 * 1 = 2.2
Laundry: 1.4 (hot and cold) = 1.4
7.2 + 1.4 + 2.2 + 1.4 = 12.2 w.s.f.u.

Step 2. Minimum daily static service pressure

System pressure is 72 psi.
Elevation difference is 6′; adjustment = 6 * (.5 psi/foot) = 3 psi.
No pressure reducing valve; therefore, no adjustment.
Two thermostatic mixing valves with a pressure loss of 6.2 psi. Since this loss is less than 8 psi, there is no loss to be adjusted.
Total losses = 3 psi.
Minimum daily static service pressure = 72 − 3 = 69 psi.

Step 3. The maximum developed length was given as 100′. If it was not given, the length must be calculated with measurements.

Step 4. Enter Table AP201.1 and use the portion of the table with a pressure range of over 60 psi. Then follow down the selected "maximum developed length" of 100′ column to a fixture unit equal to or greater than 12.2 w.s.f.u., which is 32. Then read back on that row to determine the minimum-size meter and service pipe with a ¾″ diameter and a water-service pipe of 1″ in diameter.

APPENDIX P—WATER PIPE SIZING *(cont.)*

A PORTION OF – TABLE AP201.1—CONTINUED MINIMUM SIZE OF WATER METERS, MAINS AND DISTRIBUTION PIPING BASED ON WATER SUPPLY FIXTURE UNIT (W.S.F.U.) VALUES

METER AND SERVICE PIPE (inches)	DISTRIBUTION PIPE (inches)	MAXIMUM DEVELOPMENT LENGTH (feet)									
Pressure Range Over 60		40	60	80	100	150	200	250	300	400	500
¾	½ª	3	3	3	2.5	2	1.5	1.5	1	1	0.5
¾	¾	9.5	9.5	9.5	9.5	7.5	6	5	4.5	3.5	3
¾	1	32	32	32	32	32	24	19.5	15.5	11.5	9.5
1	1	32	32	32	32	32	28	28	17	12	9.5
¾	1¼	32	32	32	32	32	32	32	32	32	30
1	1¼	80	80	80	80	80	80	69	60	46	36
1½	1¼	80	80	80	80	80	80	76	65	50	38
1	1½	87	87	87	87	87	87	87	87	87	84
1½	1½	151	151	151	151	151	151	151	144	114	94
2	1½	151	151	151	151	151	151	151	151	118	97
1	2	87	87	87	87	87	87	87	87	87	87
1½	2	275	275	275	275	275	275	275	275	275	252
2	2	365	368	368	368	368	368	368	368	318	273
2	2½	533	533	533	533	533	533	533	533	533	533

For SI: 1 inch = 25.4, 1 foot = 304.8 mm, 1 pound per square inch = 6.895 kPa.

a. Minimum size for building supply is a ¾-inch pipe.

© 2018 IRC®, International Code Council®

PLUMBING

RESIDENTIAL SPRINKLERS

1

Not required in: Attics	Not required in: Clothes closets	Not required in: Small bathroom	Not required in: Garage

1.1

1.2

1.3

1.4

1

$$P_t = P_{sup} - PL_{svc} - P_{Lm} - P_{Ld} - P_{Le} - P_{sp} \quad \text{(Equation 29-1)}$$

where:

P_t = Pressure used in applying Tables P2904.6.2(4) through P2904.6.2(9).

P_{sup} = Pressure available from the water supply source.

P_{Lsvc} = Pressure loss in the water-service pipe.

P_{Lm} = Pressure loss in the water meter.

P_{Ld} = Pressure loss from devices other than the water meter.

P_{Le} = Pressure loss associated with changes in elevation.

P_{sp} = Maximum pressure required by a sprinkler.

RESIDENTIAL SPRINKLERS *(cont.)*

No.	Code	Description
①	P2904	Sprinklers shall be installed to protect all areas of a dwelling unit except: 1. Attics, crawl spaces, and normally unoccupied concealed spaces that do not contain fuel-fired appliances 2. Clothes closets, linen closets, and pantries not exceeding 24 ft² in area, with the smallest dimension not greater than 3′ and having wall and ceiling surfaces of gypsum board 3. Bathrooms not more than 55 ft² in area 4. Garages; carports; exterior porches; unheated entry areas, such as mudrooms, that are adjacent to an exterior door; and similar areas
	P2904.6.2	Pipe shall be sized by determining the available pressure to offset friction loss in piping and identifying a piping material, diameter, and length using the equation above.
	P2904.8.1	The following items shall be verified prior to the concealment of any sprinkler system piping: 1. Sprinklers are installed in all areas as required by Section P2904.1.1 in the IRC code book. 2. Where sprinkler water spray patterns are obstructed by construction features, luminaries, or ceiling fans, additional sprinklers are installed as required by Section P2904.2.4.2 in the IRC code book. 3. Sprinklers are the correct temperature rating and are installed at or beyond the required separation distances from heat sources as required by Sections P2904.2.1 and P2904.2.2 in the IRC code book. 4. The pipe size equals or exceeds the size used in applying Tables P2904.6.2(4) through 2904.6.2(9) in the IRC code book or, if the piping system was hydraulically calculated in accordance with Section P2904.6.1 in the IRC code book, the size used in the hydraulic calculation. 5. The pipe length does not exceed the length permitted by Tables P2904.6.2(4) through P2904.6.2(9) in the IRC code book or, if the piping system was hydraulically calculated in accordance with Section P2904.6.1 in the IRC code book, pipe lengths and fittings do not exceed those used in the hydraulic calculation. 6. Nonmetallic piping that conveys water to sprinklers is listed for use with fire sprinklers. 7. Piping is supported in accordance with the pipe manufacturer's and sprinkler manufacturer's installation instructions. 8. The piping system is tested in accordance with Section P2503.7 in the IRC code book.

PLUMBING

You Should Know

- In many states, the mandatory provision has been removed. However, if sprinklers are installed, this standard is still valid for the installation.

MATERIALS, JOINTS, AND CONNECTIONS

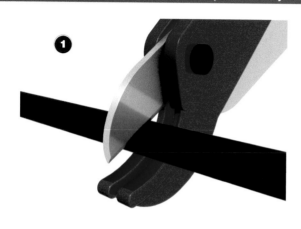

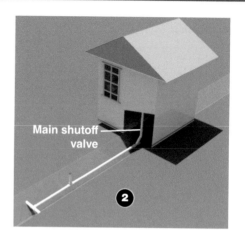

Main shutoff valve

No.	Code	Description
1	P2906.3	Polyethylene pipe shall be cut square using a cutter designed for plastic pipe. Except where joined by heat fusion, pipe ends shall be chamfered to remove sharp edges.
2	P2906.4	Water-service pipe or tubing, installed underground and outside of the structure, shall have a minimum working pressure rating of 160 psi at 73°F.
	P2906.6 and P2906.8	No saddle tap fittings are permitted and valve fittings are prohibited. Joints and connections in the plumbing system shall be gastight and water tight for the intended use or required test pressure.
	P2906.9	Joints in plastic piping shall be made with approved fittings by solvent cementing, heat fusion, corrosion-resistant metal clamps with insert fittings, or compression connections.
	P2906.10	PEX shall comply with Section P2906.9.10.
	P2906.11.1	Mechanical joints for cross-linked polyethylene/aluminum/cross-linked polyethylene shall be installed in accordance with the manufacturer's instructions.

MATERIALS, JOINTS, AND CONNECTIONS *(cont.)*

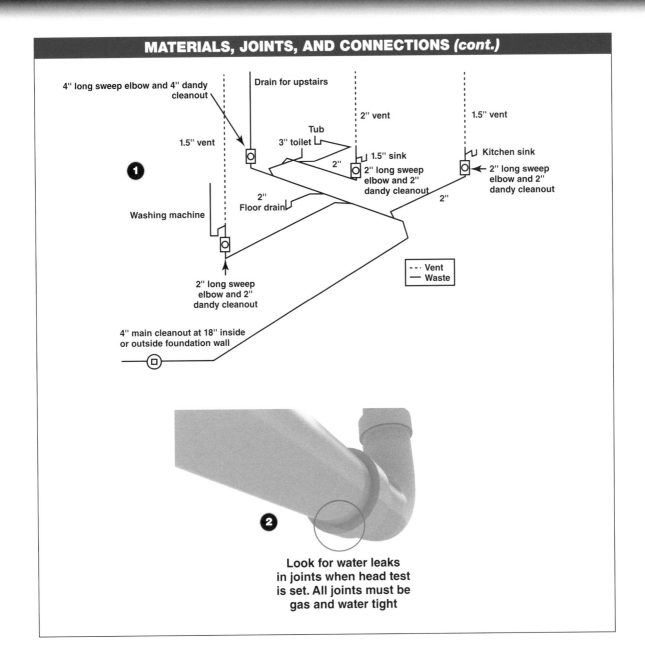

4" long sweep elbow and 4" dandy cleanout

Drain for upstairs

2" vent

1.5" vent

1.5" vent

Tub

3" toilet

1.5" sink

2"

2" long sweep elbow and 2" dandy cleanout

Kitchen sink

2" long sweep elbow and 2" dandy cleanout

Washing machine

2"
Floor drain

2"

--- Vent
— Waste

2" long sweep elbow and 2" dandy cleanout

4" main cleanout at 18" inside or outside foundation wall

Look for water leaks in joints when head test is set. All joints must be gas and water tight

PLUMBING

MATERIALS, JOINTS, AND CONNECTIONS *(cont.)*

No.	Code	Description
❶	P3001.2	No portion of the abovegrade DWV system other than vent terminals shall be located outside of a building, in attics or crawl spaces, concealed in outside walls, or in any other places subjected to freezing temperatures unless adequate provision is made to protect them from freezing by insulation or heat or both, except in localities having a winter design temperature above 32°F.
❷	P3003.1	Joints and connections in the DWV system shall be gastight and water tight for the intended use or pressure required by test.
	P3003.2	Running threads and bands shall not be used in the drainage system. Drainage and vent piping shall not be drilled, tapped, burned, or welded. The following types of joints and connections shall be prohibited: 1. Cement or concrete 2. Mastic or hot-pour bituminous joints 3. Joints made with fittings not approved for the specific installation 4. Joints between different diameter pipes made with elastomeric rolling O-rings 5. Solvent-cement joints between different types of plastic pipe, except where provided in Section P3003.13.4 6. Saddle-type fittings
	P3003.3	ABS plastic joints may be mechanical, threaded, or by solvent cement.
	P3003.9	PVC joints may be mechanical, threaded, or by solvent cement. For solvent cement, a purple primer or other approved primer may be applied unless otherwise approved. The joint must be made while the cement is still wet. Solvent cement joints are permitted both above- and belowground.

You Should Know

- There are other materials less commonly used for sanitary drainage, including cast-iron, concrete, copper pipe and tubing, steel, lead, vitrified clay, polyolefin plastic, and polyethylene plastic. Joints for these materials must be according to the manufacturer and the respective standard.

DRAINAGE FIXTURE UNITS

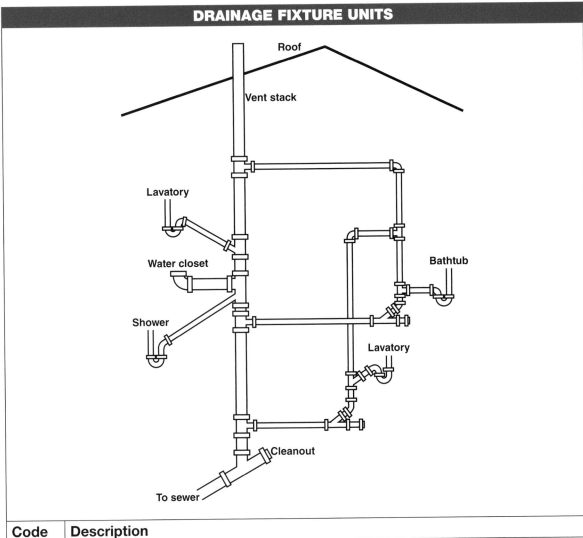

Code	Description
P3004.1	The load on DWV-system piping shall be computed in terms of drainage fixture unit (d.f.u.) values in accordance with Table P3004.1 in the IRC code book.

You Should Know

- As with water supply, drainage piping is sized based on the load for the system or each branch.
- The size of the drainage system including drain, waste, and vent pipes will all be based on this calculation.

PLUMBING

DRAINAGE FIXTURE UNITS *(cont.)*

Example

A home has two full bathroom groups, one-half bathroom group, one kitchen sink, one clothes washer, and a bar sink. All water closets have 1.6-gallon flush. Calculate the total d.f.u.

Check Table P3004.1 and find each type of fixture or group of fixtures and add the d.f.u.

1.	Full bathroom group with water closet of 1.6-gallon flush	5
2.	Full bathroom group with water closet of 1.6-gallon flush	5
3.	Half-bathroom group with water closet of 1.6-gallon flush	4
4.	Kitchen sink	2
5.	Clothes washer	2
6.	Bar sink	1
	Total:	19 d.f.u.

CHANGE OF DIRECTION AND CLEANOUTS

DRAIN PIPES—IPC 706.3 and IRC 3005.1

Fitting Type	Horizontal to Vertical	Vertical to Horizontal	Horizontal to Horizontal
Sixteenth Bend (22.5°)			
Eighth Bend (45°)			
Sixth Bend (60°)			
Quarter Bend (90°)		2" or smaller fixture drains only	2" or smaller fixture drains only
Short Sweep		2" or smaller fixture drains or 3" or larger	2" or smaller fixture drains only
Long Sweep			
Sanitary Tee			
Wye			
Combo Wye and ⅛ Bend			

Fittings must be oriented to guide the flow of drainage.

PLUMBING

CHANGE OF DIRECTION AND CLEANOUTS *(cont.)*

Building drain   Foundation wall   Cleanout ①

③   Building sewer

②

Horizontal change greater than 45 degrees requires cleanout

No.	Code	Description
①	P3005.1	Changes in direction in drainage piping shall be made by the appropriate use of sanitary tees, wyes, sweeps, and bends or by a combination of these drainage fittings in accordance with Table P3005.1 in the IRC code book.
②		Change in direction by combination fittings, heel or side inlets, or increasers shall be installed in accordance with Table P3005.1 and Sections P3005.1.1 through P3005.1.4 in the IRC code book, based on the pattern of flow created by the fitting.
	P3005.1.6	Drainage pipe sizing shall not be reduced in the direction of the flow. The following is not considered a reduction in flow: 1. A 4″ by 3″ water closet flange 2. A water closet bend fitting having a 4″ inlet and a 3″ outlet with certain conditions 3. An offset closet flange
③	P3005.2.1	Horizontal drainage pipe shall have cleanouts at not more than 100′ intervals with some exceptions.
	P3005.2.2	Building sewers smaller than 8″ must have cleanouts every 100′. Building sewers larger than 8″ must have cleanouts every 200′.
	P3005.2.3	The junction of building drain and building sewer must have cleanouts within 10′ upstream of the junction.

CHANGE OF DIRECTION AND CLEANOUTS *(cont.)*

No.	Code	Description
	P3005.2.4	Cleanouts shall be installed at each change of horizontal direction more than 45° in the building sewer, building drain, and horizontal waste or soil lines.
		Where more than one change of direction occurs in a horizontal run of piping, only one cleanout shall be required in each 40′ of developed length of the drainage piping.
	P3005.2.5	Cleanouts must be the same size as the piping served by the cleanout, except pipes larger than 4″ may have a 4″ cleanout.
	P3005.2.8	Cleanouts shall be installed so that the cleanout opens to allow cleaning in the direction of the flow of the drainage line.
	P3005.2.9	Cleanouts for 6″ and smaller piping must have 18″ clearance from and perpendicular to the face of the opening to any obstruction.
	P3005.2.10	Cleanouts shall be accessible.

You Should Know

- There shall be a cleanout near the junction of the building drain and building sewer. This cleanout shall be either inside or outside the building wall, provided that it is brought up to finish grade or to the lowest floor level. An approved two-way cleanout shall be permitted to serve as the required cleanout for both the building drain and the building sewer.

PLUMBING

SLOPE OF DRAIN PIPE

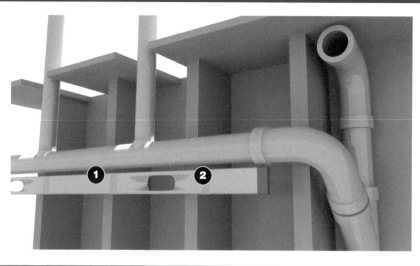

No.	Code	Description
1	P3005.3	Horizontal drainage piping shall be installed in uniform alignment at uniform slopes not less than ¼ unit vertical in 12 units horizontal (2% slope).
2		For 2½"-diameter and less, and not less than ⅛-unit vertical in 12 units horizontal (1% slope) for diameters of 3" or more.
	P3009	Subsurface landscape irrigation systems are regulated in this chapter. Materials, tests, inspections, disinfection, coloring, and sizing are all regulated, except for onsite nonpotable reuse water for subsurface landscape irrigation systems. Additionally, excavation and backfill are delineated along with distribution piping, fittings, and cleanouts.

You Should Know

- The *IRC* includes requirements for nonpotable water systems covered in Sections P2910–P2913 in the "Water Supply and Distribution" chapter. The sections include various types of nonpotable water systems.

PIPE SIZING

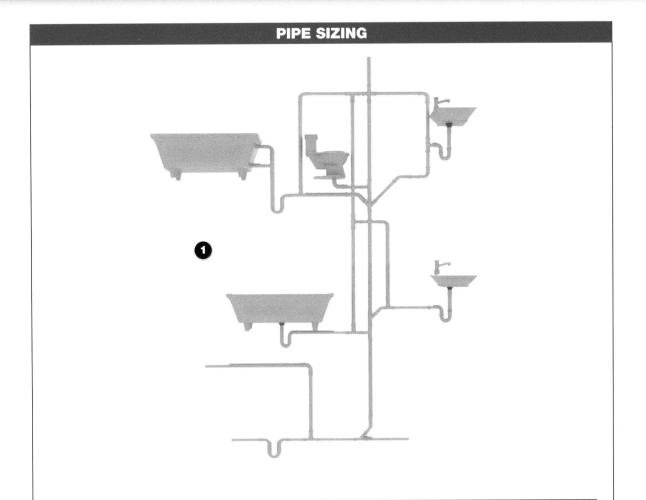

TABLE P3005.4.1 MAXIMUM FIXTURE UNITS ALLOWED TO BE CONNECTED TO BRANCHES AND STACKS

NOMINAL PIPE SIZE (inches)	ANY HORIZONTAL FIXTURE BRANCH	ANY ONE VERTICAL STACK OR DRAIN
1¼[a]	—	—
1½[b]	3	4
2[b]	6	10
2½[b]	12	20
3	20	48
4	160	240

For SI: 1 inch = 25.4 mm.

a. 1¼-inch pipe size limited to a single-fixture drain or trap arm. See Table P3201.7.
b. No water closets.

© 2018 IRC®, International Code Council®

PLUMBING

PIPE SIZING *(cont.)*

TABLE P3005.4.2 MAXIMUM NUMBER OF FIXTURE UNITS ALLOWED TO BE CONNECTED TO THE BUILDING DRAIN, BUILDING DRAIN BRANCHES, OR THE BUILDING SEWER

DIAMETER OF PIPE (inches)	SLOPE PER FOOT		
	⅛ inch	¼ inch	½ inch
1½[a, b]	—	Note a	Note a
2[b]	—	21	27
2½[b]	—	24	31
3	36	42	50
4	180	216	250

For SI: 1 inch = 25.4 mm, 1 foot = 304.8 mm.

a. 1½-inch pipe size limited to a building drain branch serving not more than two waste fixtures, or not more than one waste fixture if serving a pumped discharge fixture or garbage grinder discharge.
b. No water closets.

© 2018 IRC®, International Code Council®

No.	Code	Description
❶	P3005.4	The following general procedure is permitted to be used for sizing drainage pipe: 1. Draw an isometric layout or riser diagram denoting fixtures on the layout. 2. Assign d.f.u. values to each fixture group plus individual fixtures using Table P3004.1. 3. Starting with the top floor or most remote fixtures, work downstream toward the building drain accumulating d.f.u. values for fixture groups plus individual fixtures for each branch. Where multiple bath groups are being added, use the reduced d.f.u. values in Table P3004.1, which take into account probability factors of simultaneous use. 4. Size branches and stacks by equating the assigned d.f.u. values to pipe sizes shown in Table P3005.4.1. 5. Determine the pipe diameter and slope of the building drain and building sewer based on the accumulated d.f.u. values using Table P3005.4.2.

You Should Know

- There are two separate tables that deal with separate sanitary drain pipes; the first is for branches and stacks and the second is for building drain, building drain branches, or building sewer.

VENT TERMINATION

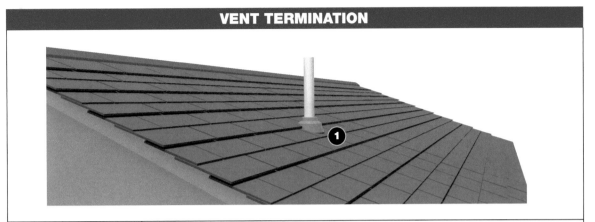

No.	Code	Description
1	P3103.1	Open vent pipes that extend through a roof shall be terminated at least 6″ above the roof or 6″ above the anticipated snow accumulation, whichever is greater. Where a roof is to be used for recreational or similar purpose, open vent pipes shall terminate not less than 7′ above the roof.
	P3103.5	An open vent terminal from a drainage system shall not be located less than 4′ directly beneath any door, openable window, or other air-intake opening of the building or of an adjacent building, nor shall any such vent terminal be within 10′ horizontally of such an opening unless it is not less than 3′ above the top of such opening.

PLUMBING

VENT CONNECTIONS AND GRADES

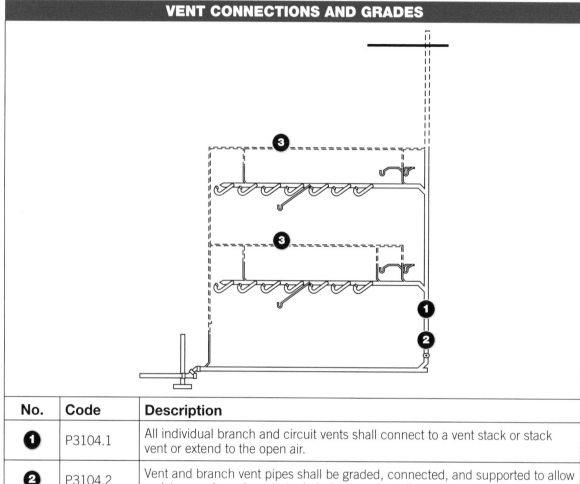

No.	Code	Description
❶	P3104.1	All individual branch and circuit vents shall connect to a vent stack or stack vent or extend to the open air.
❷	P3104.2	Vent and branch vent pipes shall be graded, connected, and supported to allow moisture and condensate to drain back to the soil or waste pipe by gravity.
	P3104.3	Every dry vent connecting to a horizontal drain shall connect above the centerline of the horizontal drain pipe.
	P3104.4	Every dry vent shall rise vertically to a minimum of 6″ above the flood level rim of the highest trap or trapped fixture being vented.
	P3104.5	A connection between a vent pipe and a vent stack or stack vent shall be made at least 6″ above the flood level rim of the highest fixture served by the vent.
❸		Horizontal vent pipes forming branch vents shall be at least 6″ above the flood level rim of the highest fixture served.

You Should Know

- Where the drainage piping has been roughed-in for future fixtures, a rough-in connection for a vent shall be installed a minimum of one-half the diameter of the drain. The vent rough-in shall connect to the vent system or shall be vented by other means. The connection shall be identified to indicate that the connection is a vent.

DRAIN, WASTE, AND VENT PIPE

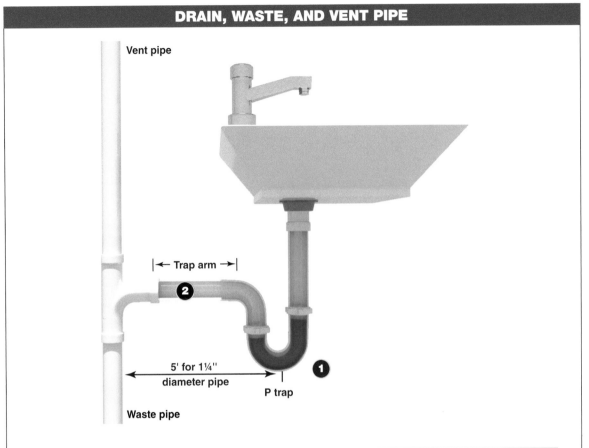

Vent pipe

Trap arm

2

5' for 1¼"
diameter pipe

P trap

1

Waste pipe

TABLE P3105.1 MAXIMUM DISTANCE OF FIXTURE TRAP FROM VENT

SIZE OF TRAP (inches)	SLOPE (inch per foot)	DISTANCE FROM TRAP (feet)
1¼	¼	5
1½	¼	6
2	¼	8
3	⅛	12
4	⅛	16
For SI: 1 inch = 25.4 mm, 1 foot = 304.8 mm, 1 inch per foot = 83.3 mm/m.		

© 2018 IRC®, International Code Council ®

PLUMBING

No.	Code	Description
1	P3105.1	Each fixture trap shall have a protecting vent located so that the slope and the developed length in the fixture drain from the trap weir to the vent fitting are within the requirements set forth in Table P3105.1.
	P3105.2	The total fall in a fixture drain resulting from pipe slope shall not exceed one pipe diameter, nor shall the vent pipe connection to a fixture drain, except for water closets, be below the weir of the trap.
2	P3105.3	A vent shall not be installed within two pipe diameters of the trap weir.

DRAIN, WASTE, AND VENT PIPE *(cont.)*

You Should Know

- The developed length of the fixture drain from the trap weir to the vent fitting for self-siphoning fixtures, such as water closets, shall not be limited.

INDIVIDUAL VENTS AND COMMON VENTS

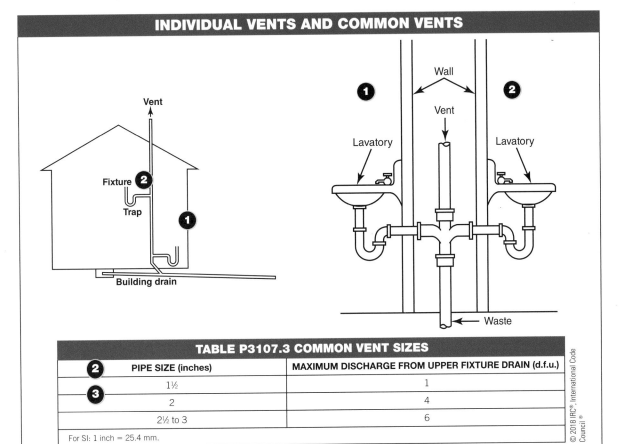

TABLE P3107.3 COMMON VENT SIZES	
PIPE SIZE (inches)	MAXIMUM DISCHARGE FROM UPPER FIXTURE DRAIN (d.f.u.)
1½	1
2	4
2½ to 3	6

For SI: 1 inch = 25.4 mm.

© 2018 IRC®, International Code Council ®

No.	Code	Description
	P3106.1	Each trap and trapped fixture is permitted to be provided with an individual vent. The individual vent shall connect to the fixture drain of the trap or trapped fixture being vented.
❶	P3107.1	An individual vent is permitted to vent two traps or trapped fixtures as a common vent. The traps or trapped fixtures being common-vented shall be located on the same floor level.
❷	P3107.2	Where the fixture drains being common-vented connect at the same level, the vent connection shall be at the interconnection of the fixture drains or downstream of the interconnection.
❸	P3107.3	Where the fixture drains connect at different levels, the vent shall connect as a vertical extension of the vertical drain. The vertical drain pipe connecting the two fixture drains shall be considered the vent for the lower fixture drain and shall be sized in accordance with Table P3107.3. The upper fixture shall not be a water closet.

You Should Know

- The distinction between a common vent and an individual vent is that a common vent serves more than one fixture. An example is a twin lavatory in a bathroom.

PLUMBING

HORIZONTAL WET VENTS

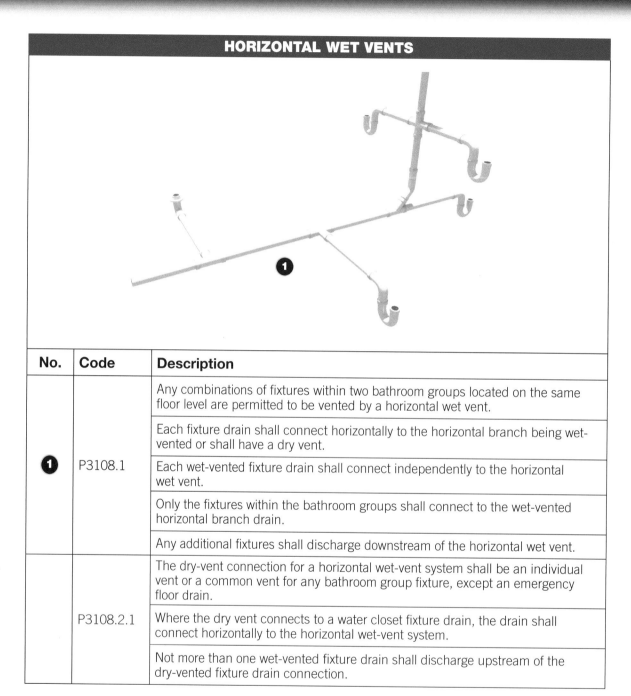

No.	Code	Description
1	P3108.1	Any combinations of fixtures within two bathroom groups located on the same floor level are permitted to be vented by a horizontal wet vent.
		Each fixture drain shall connect horizontally to the horizontal branch being wet-vented or shall have a dry vent.
		Each wet-vented fixture drain shall connect independently to the horizontal wet vent.
		Only the fixtures within the bathroom groups shall connect to the wet-vented horizontal branch drain.
		Any additional fixtures shall discharge downstream of the horizontal wet vent.
	P3108.2.1	The dry-vent connection for a horizontal wet-vent system shall be an individual vent or a common vent for any bathroom group fixture, except an emergency floor drain.
		Where the dry vent connects to a water closet fixture drain, the drain shall connect horizontally to the horizontal wet-vent system.
		Not more than one wet-vented fixture drain shall discharge upstream of the dry-vented fixture drain connection.

You Should Know

- Horizontal and vertical wet vents shall be of a minimum size as specified in Table P3108.3 in the IRC code book, based on the fixture unit discharge to the wet vent. The dry vent serving the wet vent shall be sized based on the largest required diameter of pipe within the wet-vent system served by the dry vent.

VERTICAL WET VENT

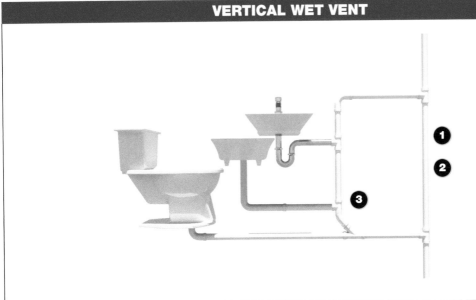

No.	Code	Description
1	P3108.2.2	The dry-vent connection for a vertical wet-vent system shall be an individual vent or common vent for the most upstream fixture drain.
2		A combination of fixtures located on the same floor level is permitted to be vented by a vertical wet vent.
3	P3108.4	The vertical wet vent shall be considered the vent for the fixtures and shall extend from the connection of the dry vent down to the lowest fixture drain connection. Each wet-vented fixture shall connect independently to the vertical wet vent. All water closet drains shall connect at the same elevation.
		Other fixture drains shall connect above or at the same elevation as the water closet fixture drains. The dry-vent connection to the vertical wet vent shall be an individual or common vent serving one or two fixtures.
	P3108.5	The maximum developed length of wet-vented fixture drains shall comply with Table P3105.1 in the IRC code book.

You Should Know

- Horizontal and vertical wet vents shall be of a minimum size as specified in Table P3108.3 in the IRC code book, based on the fixture unit discharge to the wet vent. The dry vent serving the wet vent shall be sized based on the largest required diameter of pipe within the wet-vent system served by the dry vent.

PLUMBING

ISLAND FIXTURE VENTING

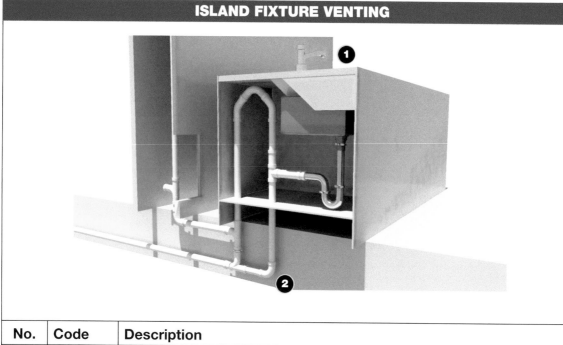

No.	Code	Description
1	P3112	Kitchen sinks with a dishwasher waste connection, a food waste grinder, or both, in combination with the kitchen sink waste, shall be permitted to be island-vented.
2	P3112.2	The island fixture vent shall connect to the fixture drain as required for an individual or common vent. The vent shall rise vertically to above the drainage outlet of the fixture being vented before offsetting horizontally or vertically downward. The vent or branch vent for multiple island fixture vents shall extend to a minimum of 6″ above the highest island fixture being vented before connecting to the outside vent terminal.
	P3112.3	The lowest point of the island fixture vent shall connect full size to the drainage system. The connection shall be to a vertical drain pipe or to the top half of a horizontal drain pipe. Cleanouts shall be provided in the island fixture vent to permit rodding of all vent piping located below the flood level rim of the fixtures.
		The vent shall be sized in accordance with Section P3113.1 in the IRC code book.

VENT PIPE SIZING

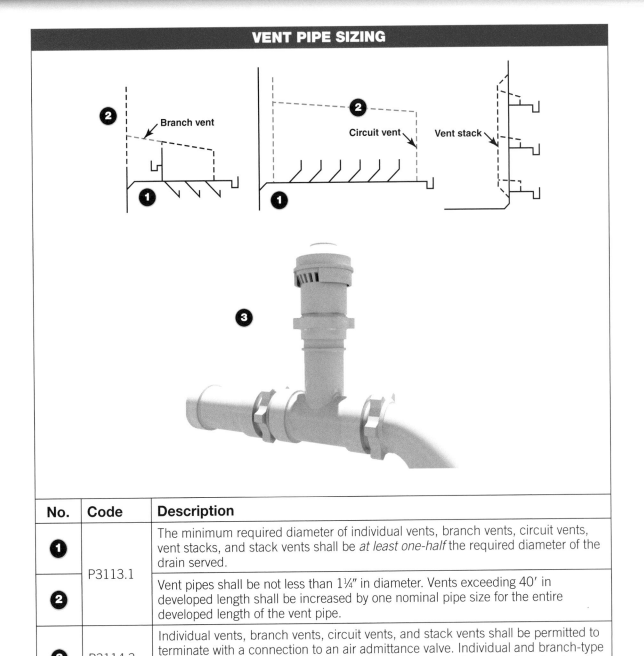

No.	Code	Description
❶	P3113.1	The minimum required diameter of individual vents, branch vents, circuit vents, vent stacks, and stack vents shall be *at least one-half* the required diameter of the drain served.
❷		Vent pipes shall be not less than 1¼″ in diameter. Vents exceeding 40′ in developed length shall be increased by one nominal pipe size for the entire developed length of the vent pipe.
❸	P3114.3	Individual vents, branch vents, circuit vents, and stack vents shall be permitted to terminate with a connection to an air admittance valve. Individual and branch-type air admittance valves shall vent only fixtures that are on the same floor level and connect to a horizontal branch drain.

PLUMBING

AIR ADMITTANCE VALVES

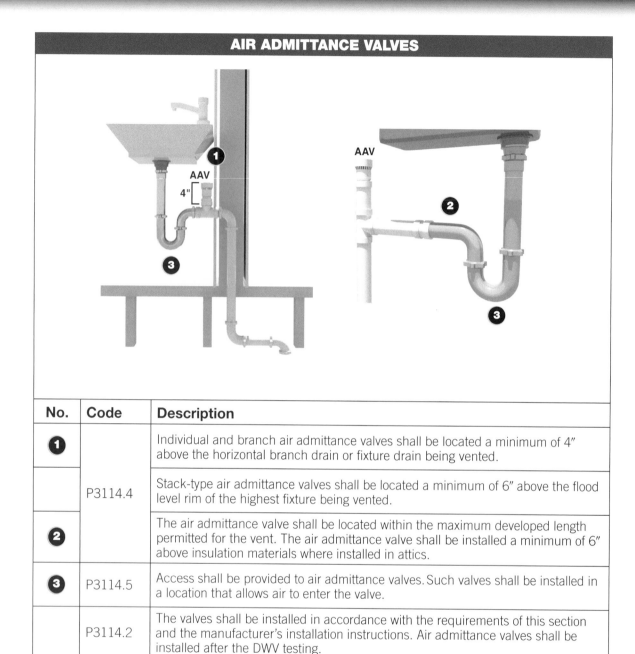

No.	Code	Description
1		Individual and branch air admittance valves shall be located a minimum of 4″ above the horizontal branch drain or fixture drain being vented.
	P3114.4	Stack-type air admittance valves shall be located a minimum of 6″ above the flood level rim of the highest fixture being vented.
2		The air admittance valve shall be located within the maximum developed length permitted for the vent. The air admittance valve shall be installed a minimum of 6″ above insulation materials where installed in attics.
3	P3114.5	Access shall be provided to air admittance valves. Such valves shall be installed in a location that allows air to enter the valve.
	P3114.2	The valves shall be installed in accordance with the requirements of this section and the manufacturer's installation instructions. Air admittance valves shall be installed after the DWV testing.

You Should Know

- Air admittance valves without an engineered design shall not be used to vent sumps or tanks of any type.

FIXTURE TRAPS

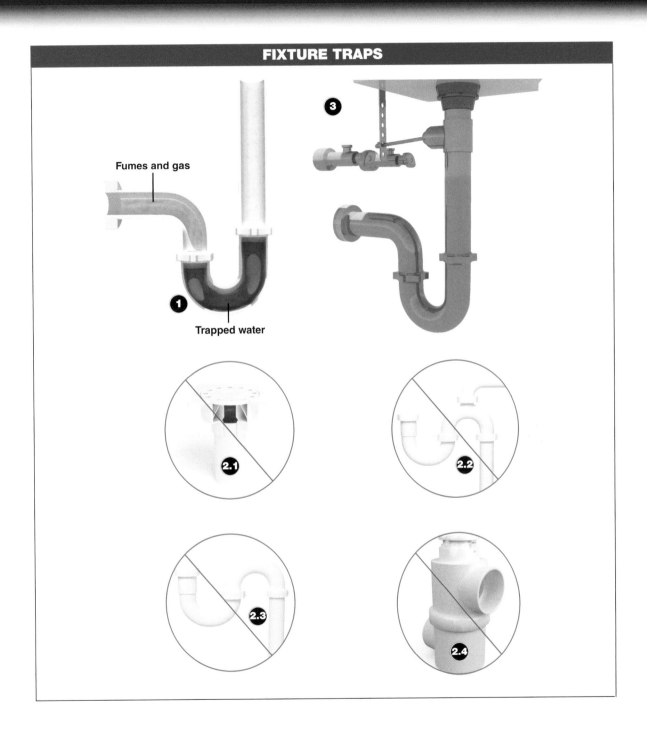

Fumes and gas

Trapped water

PLUMBING

No.	Code	Description
❶	P3201.2	Emergency floor drain traps and traps subject to evaporation shall have a liquid seal not less than 2" and not more than 4". Trap seal protection is based on type of liquid used, such as potable water, gray water, waste water, and a barrier type trap seal protection device.
❷	P3201.5	The following types of traps are prohibited: 1. Bell traps 2. Separate fixture traps with interior partitions, except those lavatory traps made of plastic, stainless-steel, or other corrosion-resistant material 3. "S" traps 4. Drum traps 5. Trap designs with moving parts
❸	P3201.6	Each plumbing fixture shall be separately trapped by a water seal trap. The vertical distance from the fixture outlet to the trap weir shall not exceed 24" and the horizontal distance shall not exceed 30" measured from the centerline of the fixture outlet to the centerline of the inlet of the trap.
		The height of a clothes washer standpipe above a trap shall conform to Section P2706.2 in the IRC code book.
	P3201.7	Fixture trap size shall be sufficient to drain the fixture rapidly and not less than the size indicated in Table P3201.7 found on page 274 of this text. A trap shall not be larger than the drainage pipe into which the trap discharges.

You Should Know

- Fixtures shall not be double trapped.

FIXTURE TRAPS *(cont.)*

TABLE P3201.7
SIZE OF TRAPS FOR PLUMBING FIXTURES

PLUMBING FIXTURE	TRAP SIZE MINIMUM (inches)
Bathtub (with or without shower head and/or whirlpool attachments)	1½
Bidet	1¼
Clothes washer standpipe	2
Dishwasher (on separate trap)	1½
Floor drain	2
Kitchen sink (one or two traps, with or without dishwasher and food waste disposer)	1½
Laundry tub (one or more compartments)	1½
Lavatory	1¼
Shower (based on the total flow rate through showerheads and body sprays) Flow rate: 5.7 gpm and less More than 5.7 gpm up to 12.3 gpm More than 12.3 gpm up to 25.8 gpm More than 25.8 gpm up to 55.6 gpm	1½ 2 3 4

For SI: 1 inch = 25.4 mm.

© 2018 IRC®, International Code Council ®

PLUMBING

SECTION 4

ELECTRICAL

WIRE CONNECTORS

Application	Description	Equipment
	Conductors shall be spliced or joined with splicing devices listed for the purpose. Splices and joints and the free ends of conductors shall be covered with an insulation equivalent to that of the conductors or with an insulating device listed for the purpose.	
	The grounding or bonding conductor shall be connected to the grounding electrode by exothermic welding, listed lugs, listed pressure connectors, listed clamps, or other listed means.	
	Cable staples are used to secure NM cable routed across or through wood framing members such as studs or plates. The staple is driven into wood framing carefully to avoid causing damage to the cable.	
Mounted in a wall (flush mounted) — Accessible only from front by opening door — Breakers used to control light, heat, or power circuits — Panelboard is housed in a cabinet	Section E3404.6 of the IRC states that unused openings, other than those intended for the operation of equipment, those intended for mounting purposes, and those permitted as part of the design for listed equipment, shall be closed to afford protection substantially equivalent to the wall of the equipment. Where metallic plugs or plates are used with nonmetallic enclosures they shall be recessed at least ¼" from the outer surface of the enclosure. Filler plates such as a knockout filler or circuit breaker knockout filler are used to fill unused holes in electrical equipment. The knockout filler is inserted into the hole with the tabs bent back to prevent withdrawal.	
	Here a circuit breaker knockout filler is used to fill unused holes in a panelboard cover. A filler is inserted and tabs prevent accidental withdrawal.	

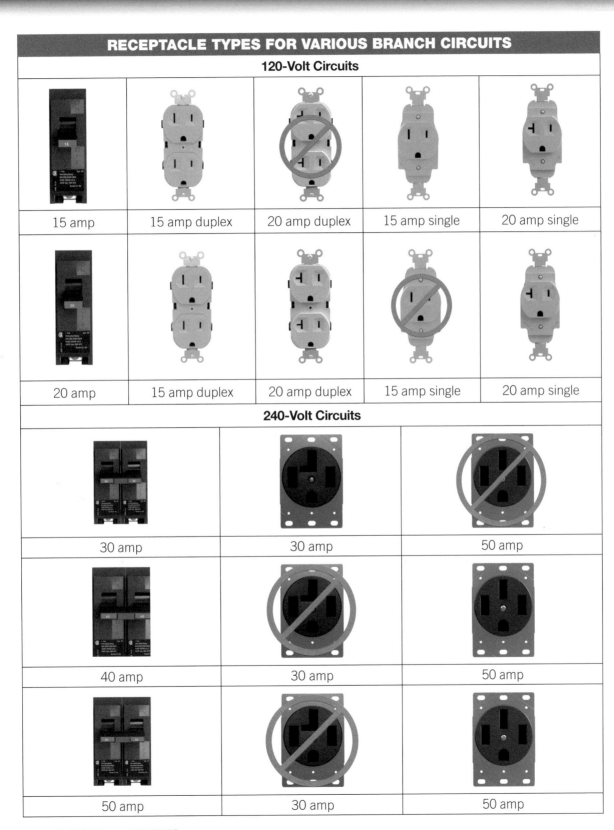

RECEPTACLE TYPES FOR VARIOUS BRANCH CIRCUITS

120-Volt Circuits

15 amp	15 amp duplex	20 amp duplex	15 amp single	20 amp single
20 amp	15 amp duplex	20 amp duplex	15 amp single	20 amp single

240-Volt Circuits

30 amp	30 amp	50 amp
40 amp	30 amp	50 amp
50 amp	30 amp	50 amp

ELECTRICAL

TYPES OF TERMINALS

Screw Terminals

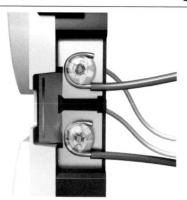

A screw terminal is a type of electrical connection where a wire is clamped down to metal plate by a screw. The wire is generally stripped of insulation at the end and is bent in a U or J shape to fit around the shaft of the screw. The wire is tightened to the device by the screw.

Push-In Terminals

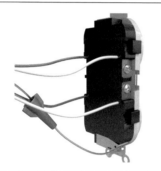

A push-in terminal slot allows an electrical conductor (normally a #14 wire) to enter the back of the receptacle through a hole slightly larger than the wire, then slide past a spring-loaded piece of metal that holds pressure against the wire. This metal pressure keeps the wire in place and from falling back out of the receptacle.

Clamping-Type Terminals

These types of terminals rely on pressure against a metal plate after inserting a conductor into a hole in the back of the receptacle. The plate is tightened with a screw connection, forcing the metal plate against the bare conductor (after removing insulation).

You Should Know

- The outlet manufacturer will specify how much insulation to remove. When removing insulated sheathing, be careful not to remove too much.

TYPES OF SWITCHES

1-Pole Switch

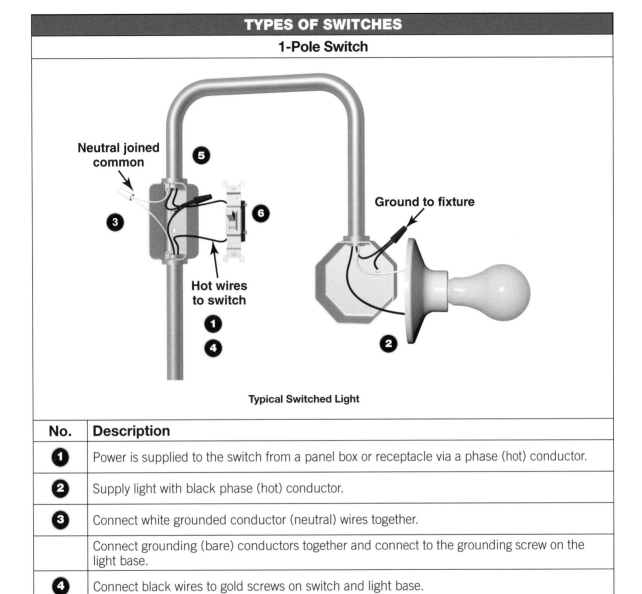

Neutral joined common

Ground to fixture

Hot wires to switch

Typical Switched Light

No.	Description
1	Power is supplied to the switch from a panel box or receptacle via a phase (hot) conductor.
2	Supply light with black phase (hot) conductor.
3	Connect white grounded conductor (neutral) wires together.
	Connect grounding (bare) conductors together and connect to the grounding screw on the light base.
4	Connect black wires to gold screws on switch and light base.
5	Connect white wire to silver screw on light base.
6	General-use and motor-circuit switches and circuit breakers shall clearly indicate whether they are in the open OFF or closed ON position. Where single-throw switches or circuit breaker handles are operated vertically rather than rotationally or horizontally, the up position of the handle shall be the ON position.

ELECTRICAL

TYPES OF SWITCHES

3-Pole Switch

Typical Three-Way Switched Light

No.	Description
1	Supply power to the first switch with two phase (hot) conductors and a ground.
2	Route the phase conductors according to the switch manufacturer's instructions. Generally, the black wire at the first switch is connected to the common. The red and black wires to the second switch connect to the two travelers of the first and second switches. The neutral wire, ground wire, and the common wire from the second switch will lead to the light base.
	Join the neutral (grounded) conductors together in both switch boxes with wire connectors.
3	Join grounding (bare) conductors to ground in each switch box and connect to the light base grounding screw.
4	General-use and motor-circuit switches and circuit breakers shall clearly indicate whether they are in the open OFF or closed ON position. Where single-throw switches or circuit breaker handles are operated vertically rather than rotationally or horizontally, the up position of the handle shall be the ON position.

RECEPTACLES

Types of Receptacles

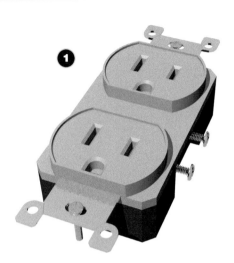

Standard Duplex Receptacle

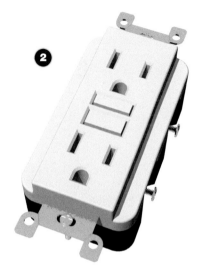

**Standard Ground-Fault
Circuit-Interrupter Receptacle**

4-Wire, 240-Volt Receptacle

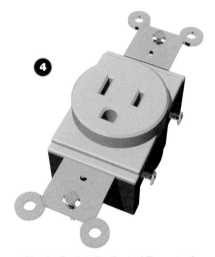

Single-Outlet, Dedicated Receptacle

No.	Description
❶	Used throughout a house as a general-use receptacle.
❷	Used in kitchens, bathrooms, and other areas.
❸	Commonly used in ranges and dryers.
❹	Limited in use to one device.

ELECTRICAL

LISTING AND LABELING

❶ TABLE E3404.4 ENCLOSURE SELECTION

PROVIDES A DEGREE OF PROTECTION AGAINST THE FOLLOWING ENVIRONMENTAL CONDITIONS	FOR OUTDOOR USE									
	Enclosure-type number									
	3	3R	3S	3X	3RX	3SX	4	4X	6	6P
Incidental contact with the enclosed equipment	X	X	X	X	X	X	X	X	X	X
Rain, snow, and sleet	X	X	X	X	X	X	X	X	X	X
Sleet[a]	—	—	X	—	—	X	—	—	—	—
Windblown dust	X	—	X	X	—	X	X	X	X	X
Hosedown	—	—	—	—	—	—	X	X	X	X
Corrosive agents	—	—	—	X	X	X	—	X	—	X
Temporary submersion	—	—	—	—	—	—	—	—	X	X
Prolonged submersion	—	—	—	—	—	—	—	—	—	X

PROVIDES A DEGREE OF PROTECTION AGAINST THE FOLLOWING ENVIRONMENTAL CONDITIONS	FOR INDOOR USE									
	Enclosure-type number									
	1	2	4	4X	5	6	6P	12	12K	13
Incidental contact with the enclosed equipment	X	X	X	X	X	X	X	X	X	X
Falling dirt	X	X	X	X	X	X	X	X	X	X
Falling liquids and light splashing	—	X	X	X	X	X	X	X	X	X
Circulating dust, lint, fibers and flyings	—	—	X	X	—	X	X	X	X	X
Settling airborne dust, lint, fibers and flings	—	—	X	X	X	X	X	X	X	X
Hosedown and splashing water	—	—	X	X	—	X	X	—	—	—
Oil and coolant seepage	—	—	—	—	—	—	—	X	X	X
Oil or coolant spraying and splashing	—	—	—	—	—	—	—	—	—	X
Corrosive agents	—	—	—	X	—	—	X	—	—	—
Temporary submersion	—	—	—	—	—	X	X	—	—	—
Prolonged submersion	—	—	—	—	—	—	X	—	—	—

a. Mechanism shall be operable when ice covered.

Note 1: The term raintight is typically used in conjunction with Enclosure Types 3, 3S, 3SX, 3X, 4, 4X, 6, and 6P. The term rainproof is typically used in conjunction with Enclosure Types 3R and 3RX. The term water tight is typically used in conjunction with Enclosure Types 4, 4X, 6, and 6P. The term driptight is typically used in conjunction with Enclosure Types 2, 5, 12, 12K, and 13. The term dusttight is typically used in conjunction with Enclosure Types 3, 3S, 3SX, 3X, 5, 12, 12K, and 13.

Note 2: Ingress protection (IP) ratings are found in ANSI/NEMA 60529, *Degrees of Protection Provided by Enclosures.* IP ratings are not a substitute for enclosure-type ratings.

© 2018 IRC®, International Code Council®

LISTING AND LABELING *(cont.)*

No.	Code	Description
	E3403.3	All electrical parts must be listed and bear the label of an approved listing agency and labeled so as to identify them as appropriate for the installation as intended by the manufacturer. They must be installed, used, or both in accordance with any instructions included in the listing and labeling.
1	E3404.4	Electrical enclosures rated less than 600 volts, installed in the proper location must be marked with enclosure type that corresponds with Table E3404.4. You can use this table for selecting the proper enclosure for the type of use, whether indoor or outdoor.
	E3405.1	All electrical equipment must have access and working space as shown in Figure E3405.1. Generally, there must be 6'6" in height, 30" in width, and 36" in depth in front of the equipment.
	E3405.7	All electrical equipment must be illuminated to service equipment. This can be a switched light. Daylight by itself is not sufficient for required illumination.

You Should Know

- Labeled equipment or materials are those having an attached label, symbol, or other identifying mark of an organization acceptable to the authority having jurisdiction and concerned with product evaluation that maintains periodic inspection of production of labeled equipment or materials and by whose labeling the manufacturer indicates compliance with appropriate standards or performance in a specified manner.

- Listed equipment, materials, or services are those included in a list published by an organization that is acceptable to the authority having jurisdiction and concerned with evaluation of products or services, that maintains periodic inspection of production of listed equipment or materials or periodic evaluation of services, and whose listing states either that the equipment, material, or services meets identified standards or has been tested and found suitable for a specified purpose.

ELECTRICAL

CONDUCTORS: CONNECTIONS AND IDENTIFICATION

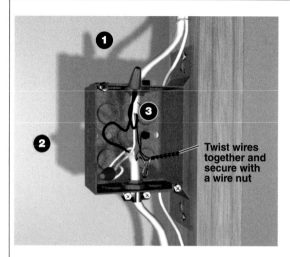

Twist wires together and secure with a wire nut

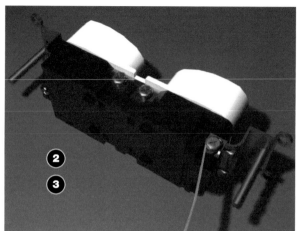

No.	Code	Description
1	E3406.8	Where wires are joined together, the terminals and splices between conductors must be the appropriate type for the material. Generally, dissimilar materials such as copper and aluminum are not permitted to be joined together unless the connecting device is listed for this purpose.
2	E3406.11.3	Where conductors join fixtures or devices, a minimum length of 6″ free conductor must be provided at each outlet, junction, or switch to ensure that the connection can be removed or replaced safely. Where that opening to an outlet is less than 8″ in any dimension, at least 3″ of the conductor must extend outside the opening.
3	E3407.1	Number 6 AWG or smaller insulated grounded conductors must be identified with a continuous white or gray color. Another option is three continuous white stripes over any color besides green.
	E3407.2	Number 6 AWG equipment grounding conductors must be identified by a continuous green color or one or more yellow stripes over continuous green. A grounding conductor may also be bare wire.

You Should Know

- Insulation on the ungrounded conductors shall be a continuous color other than white, gray, and green.
- Section E3406.13 specifies eight methods for connection of grounding and bonding equipment. You must use one or more of these methods for the connection.
- Where a tightening torque is indicated as a numeric value on equipment or installation instructions, a calibrated torque tool shall be used to achieve the indicated torque value per Section E3406.12.

DEFINITIONS AND TERMS	
Readily Accessible	Capable of being reached quickly for operation, renewal, or inspections, without requiring those to whom ready access is requisite to climb over or remove obstacles or to resort to portable ladders, etc.
Arc-Fault Circuit Interrupter	A device intended to provide protection from the effects of arc faults by recognizing characteristics unique to arcing and by functioning to de-energize the circuit when an arc fault is detected.
Branch Circuit	The circuit conductors between the final overcurrent device protecting the circuit and the outlet(s).
Device	A unit of an electrical system that carries or controls electrical energy as its principal function.
Grounded/ Grounding	Connected (connecting) to ground or to a conductive body that extends the ground connection.
Grounded Conductor	A system or circuit conductor that is intentionally grounded.
Equipment Grounding Conductor (EGC)	The conductive path installed to connect normally noncurrent-carrying metal parts of equipment together and to the system grounded conductor, the grounding electrode conductor, or both.
Grounding Electrode	A conducting object through which a direct connection to earth is established.
Grounding Electrode Conductor	A conductor used to connect the system grounded conductor or the equipment to a grounding electrode or to a point on the grounding electrode system.
Ground-Fault Circuit-Interrupter	A device intended for the protection of personnel that functions to de-energize a circuit or portion thereof within an established period of time when a current to ground exceeds the value for a Class A device.
Intersystem Bonding Termination	A device that provides a means for connecting a communications system(s), grounding conductor(s), and bonding conductor(s) at the service equipment or at the disconnecting means for buildings or structures supplied by a feeder or branch.
Listed	Equipment, materials, or services included in a list published by an organization that is acceptable to the authority having jurisdiction and concerned with evaluation of products or services, that maintains periodic inspection of production of listed equipment or materials or periodic evaluation of services, and whose listing states either that the equipment, material, or services meets identified standards or has been tested and found suitable for a specified purpose.
Qualified Person	One who has the skills and knowledge related to the construction and operation of the electrical equipment and installations and has received safety training to recognize and avoid the hazards involved.
Service and Service Drop	Service is the conductors and equipment for delivering energy from the serving utility to the wiring system of the premises served. Service drop is the overhead service conductors from the last pole or other aerial support to and including the splices, if any, connecting to the service-entrance conductors at the building or other structure.

ELECTRICAL

GENERAL

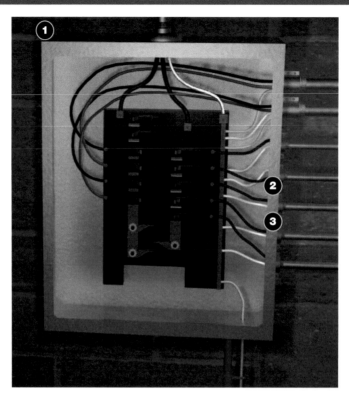

No.	Code	Description
❶	E3601.2	Any one- or two-family dwelling cannot have more than one service.
❷	E3601.6	You must be able to disconnect all conductors in a building or structure from the service entrance conductors.
❸	E3601.6.1	A service disconnect must be clearly and permanently marked as a disconnecting means.
	E3601.6.2	This service disconnect must be readily accessible either outside or inside near where the service enters the building. It cannot be in bathrooms. The occupants must have access in case they need to quickly disconnect the service of the dwelling unit in which they live.
	E3601.7	The service disconnect must have no more than six switches or six sets of circuit breakers mounted in one enclosure or a group of separate enclosures.

You Should Know

- Service conductors supplying a building or other structure shall *not* pass through the interior of another building or other structure.

SIZE AND RATING

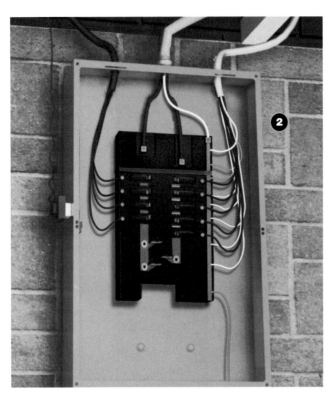

No.	Code	Description
❶	E3602.1	Ungrounded service conductors must be rated for at least the ampacity of the load served. For one-family dwellings, at least a 3 wire, 100 amperes service is a minimum. For all other uses, 60 amperes is the minimum.
❷	E3602.2	The minimum load for ungrounded service conductors and devices that serve all of the dwelling must be derived in accordance with Table E3602.2 in the IRC code book. Ungrounded service conductors that serve all of the dwelling must be sized as feeders according to Chapter 37 in the IRC code book.
	E3602.2.1	Services not required to be 100 amperes (see Table E3602.1 in the IRC code book) must be sized according to Chapter 37 in the IRC code book.
	E3602.3	Combined rating for all service disconnects serving a one-family dwelling must be at least the load calculated by using Table E3602.2 in the IRC code book.
	E3602.4	Systems shall be three-wire, $^{120}/_{240}$-volt, single-phase with a grounded neutral.

ELECTRICAL

SIZE AND RATING

MINIMUM SERVICE LOAD CALCULATION

Given: 1740 ft^2 detached single-family home with the following appliances

Appliance	Nameplate Rating
(a) Range/Oven	8000 watts
(b) Clothes Dryer	4500 watts
(c) Water Heater	4500 watts
(d) Heat Pump	12,000 watts (no supplemental heating)
(e) Two Appliance Circuits	3000 watts

Steps

1	1740 * (3 watts/ ft^2) = 5220 watts
2	3 appliance circuits and 1 laundry circuit = (1500) = 4500 watts
3	Range/Oven = 8000 watts
4	Clothes Dryer = 4500 watts
5	Water Heater = 4500 watts
6	Total load = 38,720 watts
7	100% of the first 10,000 = 10,000
8	40% of 28,720 = 11,488
9	Subtotal = 21,488 watts
10	+100% of heat pump = 12,000 watts = 33,488 watts
11	Divide by system voltage: $^{33,488}/_{240}$ = 139.5 amps

You Should Know

- The combined rating of all individual service disconnects serving a single dwelling unit shall not be less than the load determined from Table E3602.2 in the IRC code book and shall not be less than as specified in Section E3602.1 in the IRC code book.

GROUNDED AND UNGROUNDED SERVICE CONDUCTOR SIZE

① TABLE E3603.4
GROUNDING ELECTRODE CONDUCTOR SIZE[a, b, c, d, e, f]

SIZE OF LARGEST UNGROUNDED SERVICE-ENTRANCE CONDUCTOR OR EQUIVALENT AREA FOR PARALLEL CONDUCTORS (AWG/kcmil)		SIZE OF GROUNDING ELECTRODE CONDUCTOR (AWG/kcmil)	
Copper	Aluminum or copper-clad aluminum	Copper	Aluminum or copper-clad aluminum
2 or smaller	1/0 or smaller	8	6
1 or 1/0	2/0 or 3/0	6	4
2/0 or 3/0	4/0 or 250	4	2
Over 3/0 through 350	Over 250 through 500	2	1/0
Over 350 through 600	Over 500 through 900	1/0	3/0

© 2018 IRC®, International Code Council®

a. If multiple sets of service-entrance conductors connect directly to a service drop, set of overhead service conductors, set of underground service conductors, or service lateral, the equivalent size of the largest service-entrance conductor shall be determined by the largest sum of the areas of the corresponding conductors of each set.

b. Where there are no service-entrance conductors, the grounding electrode conductor size shall be determined by the equivalent size of the largest service-entrance conductor required for the load to be served.

c. Where protected by a ferrous metal raceway, grounding electrode conductors shall be electrically bonded to the ferrous metal raceway at both ends. [250.64(E)(1)]

d. An 8 AWG grounding electrode conductor shall be protected with rigid metal conduit, intermediate metal conduit, rigid polyvinyl chloride (Type PVC) nonmetallic conduit, rigid thermosetting resin (Type RTRC) nonmetallic conduit, electrical metallic tubing or cable armor. [250.64(B)]

e. Where not protected, 6 AWG grounding electrode conductor shall closely follow a structural surface for physical protection. The supports shall be spaced not more than 24 inches on center and shall be within 12 inches of any enclosure or termination. [250.64(B)]

f. Where the sole grounding electrode system is a ground rod or pipe as covered in Section E3608.3, the grounding electrode conductor shall not be required to be larger than 6 AWG copper or 4 AWG aluminum. Where the sole grounding electrode system is the footing steel as covered in Section E3608.1.2, the grounding electrode conductor shall not be required to be larger than 4 AWG copper conductor. [250.66(A) and (B)]

No.	Code	Description
①	E3603.1	Service and feeder conductor supplied by a single phase 120/240-volt system shall be sized according to E3603.1.1 through E3603.1.4 and Table E3705.1
	E3603.2	Ungrounded service conductors not in dwelling units must have an ampacity not less than 60 amperes and sized according to Chapter 37 in the IRC code book.

GROUNDED AND UNGROUNDED SERVICE CONDUCTOR SIZE *(cont.)*		
No.	**Code**	**Description**
	E3603.3.1	Overload protection must be an overcurrent device installed in series with each ungrounded service conductor. It must have a rating of at least that determined by Table E3603.1. This can be a set of fuses if they protect all of the ungrounded conductors of a circuit. Properly grouped single-pole circuit breakers are considered as one protective device.

You Should Know

- The feeder conductors to a dwelling unit shall not be required to have an allowable ampacity greater than that of the service-entrance conductors that supply them.
- The grounding electrode conductors shall be sized based on the size of the service entrance conductors as required in Table E3603.1.

LISTING AND LABELING

Figure E3604.1
Clearances From Building Openings
For SI: 1 foot = 304.8 mm

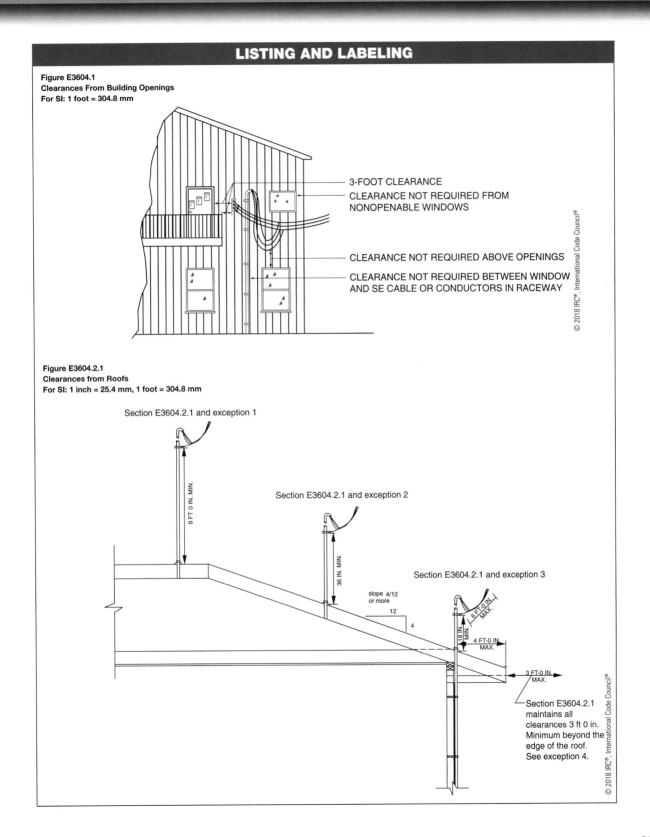

3-FOOT CLEARANCE

CLEARANCE NOT REQUIRED FROM
NONOPENABLE WINDOWS

CLEARANCE NOT REQUIRED ABOVE OPENINGS

CLEARANCE NOT REQUIRED BETWEEN WINDOW
AND SE CABLE OR CONDUCTORS IN RACEWAY

© 2018 IRC®, International Code Council®

Figure E3604.2.1
Clearances from Roofs
For SI: 1 inch = 25.4 mm, 1 foot = 304.8 mm

Section E3604.2.1 and exception 1

8 FT 0 IN. MIN.

Section E3604.2.1 and exception 2

36 IN. MIN.

Section E3604.2.1 and exception 3

slope 4/12
or more

12

4

6 FT-0 IN. MAX.

18 IN. MIN.

4 FT-0 IN. MAX.

3 FT-0 IN. MAX.

Section E3604.2.1
maintains all
clearances 3 ft 0 in.
Minimum beyond the
edge of the roof.
See exception 4.

© 2018 IRC®, International Code Council®

ELECTRICAL

LISTING AND LABELING *(cont.)*	
Code	**Description**
E3604.2.1	All conductors must have a vertical clearance of at least 8' above the roof surface and at least 3' from the edge of the roof in all directions except for the following five exceptions: 1. Those above the roof where pedestrian traffic is beneath must meet requirements of Section E3604.2.2 in the IRC code book. 2. Where the roof is sloped 4:12 or greater can have a reduced clearance of 3'. 3. Conductors above an overhang can have a reduced clearance of 18" where not more than 6' of the conductor passes over 4' or less of the roof surface and the conductors terminate at a through-the-roof raceway. 4. Maintaining the vertical clearance for a distance of 3' from the edge of roof is not necessary for the final conductor at the service drop on the side of a building. 5. Where the voltage between conductors is not above 300 volts, the clearance may be reduced to 3' if the area is guarded or isolated.
E3604.2.2	Service-drop conductors must be at least: 1. Ten feet above pedestrian access such as sidewalks or above final grade at the electrical service entrance at the lowest point of the drip loop. 2. Twelve feet above residential property and driveways. 3. Eighteen feet above public streets, alleys, roads, or parking areas where there may be truck traffic.

You Should Know

- Open overhead service conductors and multiconductor cables without an overall outer jacket shall have a clearance of not less than 3' from the sides of doors, porches, decks, stairs, ladders, fire escapes, and balconies, and from the sides and bottom of windows that open. See Figure E3604.1.

SERVICE-ENTRANCE CONDUCTORS

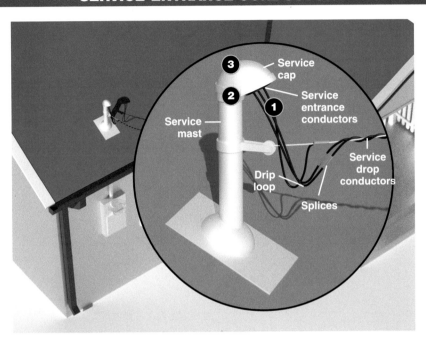

No.	Code	Description
❶	E3605.1	Service-entrance conductors that carry current must be insulated.
	E3605.3	Service-entrance conductors may be spliced or tapped but must be inside an enclosure or, if buried underground, with a listed splice.
	E3605.5	Where service-entrance cables are above ground and subject to physical damage, protection is required by rigid metal conduit, intermediate conduit, Schedule 80 PVC conduit, electrical metal tubing, reinforced thermosetting resin conduit, or other approved methods.
❷	E3605.9.1	Service raceways must have a rain-tight service head where connected to service drop conductors.
❸	E3605.9.2	Overhead service entrance cables must have a service head or an approved gooseneck.

You Should Know

- Service-drop conductors and service-entrance conductors shall be arranged so that water will not enter service raceways or equipment.

ELECTRICAL

SERVICE EQUIPMENT AND SYSTEM GROUNDING

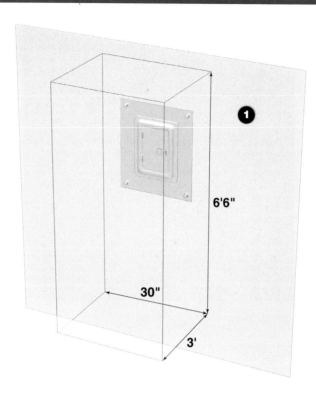

No.	Code	Description
	E3606.1	Any electrical part of the service that is energized must be enclosed.
	E3606.3	Service equipment must have a fault current rating of at least 10,000 amperes and be approved for the intended use.
	E3606.4	Service equipment must be marked as approved for the intended use as service equipment. Individual meter socket enclosures are not considered service equipment but shall be rated for the voltage and ampacity of the service.
❶	E3606.2	The working space for service equipment must be at least that specified in Table E3603.1 in the IRC code book.
	E3607.1	The premise wiring system must be grounded at the service. The ground must be sized according to Figure E3405.1 in the IRC code book.

You Should Know

- Grounding electrode conductors shall be sized in accordance with Table E3603.1 in the IRC code book.

SUBPANELS

No.	Code	Description
	E3706.2	All circuits must be legibly marked to identify them for the use intended.
❶		Spare circuit breaker positions that contained unused devices must be marked as a spare.
❷		The circuit breaker identification must be on the face of the panelboard enclosure.
❸		Identification must be distinct and not based on transient condition such as Bobby's bedroom or Sarah's workshop.
	E3706.4	Each grounded conductor must terminate inside the panelboard on a single terminal (do not add more than one wire per terminal). The only exception to this is more than one parallel grounded conductors may connect to a common terminal.

You Should Know

- Subpanels located at detached buildings or structures supplied by feeder(s) or branch circuit(s) from another building shall have a grounding electrode or grounding electrode system installed in accordance with Section E3608 in the IRC code book.
- The equipment grounding bar must be bonded to the subpanel enclosure.
- A main breaker is not needed if there is overcurrent protection.

ELECTRICAL

GROUNDING ELECTRODE SYSTEM

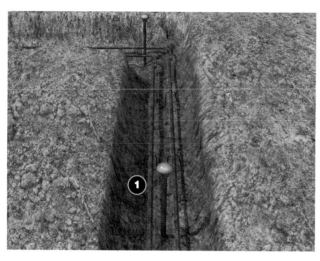

No.	Code	Description
1	E3608.1.2	A concrete encased electrode at least 20' long that is either: 1. One or more steel reinforcing bars at least ½" in diameter and 20' long installed in one continuous length. 2. A bare copper wire that is #4 AWG or larger. These metallic components are considered a grounding electrode as long as the metal is encased in at least 2" of concrete in a foundation or footing in direct contact with earth. This is known as a Ufer Ground.
	E3608.1.3	Another grounding electrode is a ground ring at least #2 AWG bare copper conductor around a building and at least 20' long, buried at least 30" in direct contact with earth.

You Should Know

- A single electrode consisting of a rod, pipe, or plate that does not have a resistance to ground of 25 ohms or less shall be augmented by one additional electrode of any of the types specified in Sections E3608.1.2 through E3608.1.6 in the IRC code book. Where multiple listed electrodes or rod, pipe, or plate electrodes are installed to meet the requirements of this section, they shall be not less than 6' apart.

METALLIC WATER PIPE AND GROUND RODS

No.	Code	Description
METALLIC WATER PIPE AND GROUND RODS *(cont.)*		
1	E3608.1.1	Another grounding electrode is a continuous metal underground water pipe in contact with earth at least 10' in length. This includes well casings. Use of a metal underground pipe must be supplemented with an additional grounding electrode.
		Interior metal pipe more than 5' from its entrance to the building is not to be used as a conductor to interconnect electrodes of the grounding electrode system.
2	E3608.1.4	Another grounding electrode is a rod and pipe not less than 8' in length and: 1. Pipe or conduit not smaller than ¾" and must be made of iron or steel that is coated to protect from corrosion. 2. Rod-type grounding electrodes that are made of stainless steel, copper, or zinc-coated steel at least ⅝" in diameter.
	E3608.4	Any grounding system using a single rod, pipe, or plate electrode must include an additional electrode. This supplemental electrode must be bonded to: 1. A rod, pipe, or plate electrode, or 2. A grounding electrode conductor, or 3. A grounded service entrance conductor, or 4. A nonflexible grounded service raceway, or 5. A grounded service enclosure.

You Should Know

- When multiple rods, pipes, or plate electrodes are installed, they must be at least 6' apart from each other.

BONDING

No.	Code	Description
1	E3609.1	Bonding is required where needed to ensure electrical continuity and the ability to carry the load of any fault current to a path to grounding.
2	E3609.2	All noncurrent-carrying metal parts must be adequately bonded together including: 1. Raceways or service cable armor or sheath that has service conductors, and 2. Service enclosures with service conductors.
3	E3609.3	A properly sized intersystem bonding termination (IBT) fitting is required for other systems such as telephone or cable television outside enclosures at disconnecting means of the service equipment. An aluminum or copper bus bar not less than ¼" by 2" wide and long enough to accommodate at least three terminations for communications systems shall be provided per Section E3609.3.2.
	E3609.6	Metal water pipe must be bonded to the service equipment enclosure.
4	E3609.7	Metal piping systems in or attached to a building that are capable of being energized must be bonded to the service equipment enclosure, the grounded conductor at the service, sufficiently sized grounding electrode conductor, or to the grounding electrodes.

You Should Know

- The bonding jumper shall be sized not smaller than that permitted by Table E3603.4 in the IRC code book per Section E3609.5.
- The points of attachment of the bonding jumper(s) shall be accessible.

ELECTRICAL

GROUNDING ELECTRODE CONDUCTORS AND CONNECTIONS

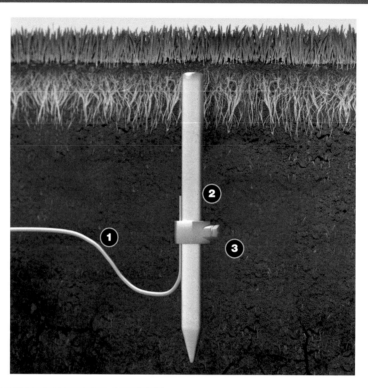

No.	Code	Description
❶	E3610.1	A grounding electrode conductor must be one continuous length and connected to a grounding electrode. Other electrodes must be connected to bonding jumpers in accordance with Section E3608.2 in the IRC code book or to one or more grounding electrodes.
❷	E3611.1	Grounding or bonding conductors must be connected to the grounding conductor with an exothermic weld, listed lugs, clamps, or pressure connectors or similar method.
❸	E3611.2	All mechanical connections to terminate a grounding electrode or bonding jumper to the grounding electrodes, which are not within concrete, must be accessible.
	E3611.7	Paint or similar nonconductive covering on equipment required to be grounded must be removed from threads and other contact surfaces to ensure electrical continuity.

You Should Know

- Connections depending on solder shall not be used.

BRANCH CIRCUIT RATINGS

TABLE E3702.14 BRANCH-CIRCUIT REQUIREMENTS-SUMMARY[a, b]

	❶ CIRCUIT RATING		
	15 amp ❷	20 amp ❷	30 amp ❸
Conductors: Minimum size (AWG) circuit conductors	14	12	10
Maximum overcurrent protection device rating ampere rating	15	20	30
Outlet devices: Lampholders permitted receptacle rating (amperes)	Any type 15 maximum	Any type 15 or 20	N/A 30
Maximum load (amperes)	15	20	30

a. These gages are for copper conductors.
b. N/A means not allowed.

© 2018 IRC®, International Code Council®

No.	Code	Description
❶	E3702.2	Branch circuits must be rated according to the maximum allowable ampere rating of the circuit breaker (or other overcurrent device). The options are 15, 20, 30, 40, or 50 amperes.
❷	E3702.3	A 15- or 20-ampere branch circuit can supply lighting units or utilization equipment or both.
❸	E3702.4	A 30-ampere branch circuit can serve fixed utilization equipment. Any cord and plug equipment cannot exceed 80% of the branch circuit amperage rating.

You Should Know

- The requirements for circuits having two or more outlets, or receptacles, other than the receptacle circuits of Sections E3703.2 and E3703.3, are summarized in Table E3702.13.

ELECTRICAL

REQUIRED BRANCH CIRCUITS

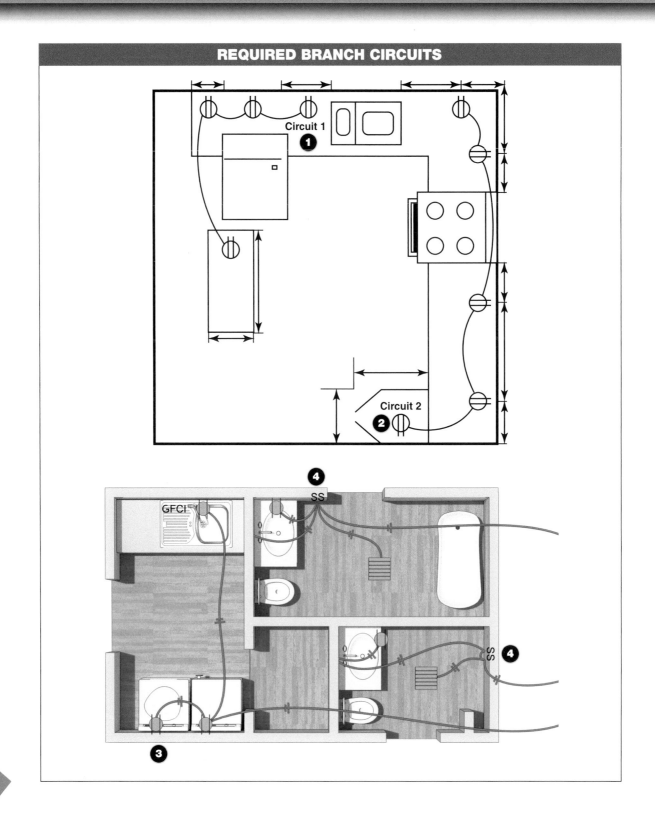

REQUIRED BRANCH CIRCUITS *(cont.)*

No.	Code	Description
	E3703.1	Except for fixed electrical space heating, central heating equipment must have an individual branch circuit. Permanent air conditioning equipment directly part of central heating equipment, such as pumps and motors, may connect to the same branch circuit.
❶	E3703.2	At least two 20-ampere branch circuits must be provided in wall and floor receptacles in kitchen, pantry, breakfast area, and dining area.
❷		Outlets that serve kitchen countertops must be served by at least two 20-ampere branch circuits. These may serve other receptacle outlets in the same kitchen, pantry, breakfast area, and dining area.
❸	E3703.3	At least one 20-ampere branch circuit is required in the laundry area. This branch circuit may not serve any other uses.
❹	E3703.4	At least one 20-ampere branch circuit is required for outlets in bathroom(s). This branch circuit may not serve any other uses.

You Should Know

- The receptacle outlet for refrigeration appliances shall be permitted to be supplied from an individual branch circuit rated 15 amperes or greater.

ELECTRICAL

CONDUCTOR SIZING AND OVERCURRENT PROTECTION

TABLE E3705.1 ALLOWABLE AMPACITIES

CONDUCTOR SIZE	CONDUCTOR TEMPERATURE RATING						CONDUCTOR SIZE
	60°C	75°C	90°C	60°C	75°C	90°C	
AWG kcmil	Types TW, UF	Types RHW, THHW, THW, THWN, USE, XHHW	Types RHW-2, THHN, THHW, THW-2, THWN-2, XHHW, XHHW-2, USE-2	Types TW, UF	Types RHW, THHW, THW, THWN, USE, XHHW	Types RHW-2, THHN, THHW, THW-2, THWN-2, XHHW, XHHW-2, USE-2	AWG kcmil
	Copper			Aluminum or copper-clad aluminum			
14[a]	15	20	25	—	—	—	—
12[a]	20	25	30	15	20	25	12[a]
10[a]	30	35	40	25	30	35	10[a]
8	40	50	55	35	40	45	8
6	55	65	75	40	50	55	6
4	70	85	95	55	65	75	4
3	85	100	115	65	75	85	3
2	95	115	130	75	90	100	2
1	110	130	145	85	100	115	1
1/0	125	150	170	100	120	135	1/0
2/0	145	175	195	115	135	150	2/0
3/0	165	200	225	130	155	175	3/0
4/0	195	230	260	150	180	205	4/0
250	215	255	290	170	205	230	250
300	240	285	320	195	230	260	300
350	260	310	350	210	250	280	350
400	280	335	380	225	270	305	400
500	320	380	430	260	310	350	500
600	350	420	475	285	340	385	600
700	385	460	520	315	375	425	700
750	400	475	535	320	385	435	750
800	410	490	555	330	395	445	800
900	435	520	585	355	425	480	900

For SI: °C = [(°F) − 32]/1.8.

a. See Table E3705.5.3 for conductor overcurrent protection limitations.

© 2018 IRC®, International Code Council®

Code	Description
E3705.1	Allowable ampacities for conductors must be determined by using Table E3705.1 along with correction and adjustment factors found in Section E3705 in the IRC code book.
E3705.2	When the ambient temperature is not 30°, determine the adjustment factor in Table E3705.2, then multiply the allowable ampacity in the table by this adjustment factor.
E3705.3	Where more than three current carrying conductors are within a raceway of cable or where conductor cables are stacked or bundled in runs of more than 2′, determine the de-rating in Table E3705.3, then multiply this by the allowable ampacity in Table E3705.1. See four exceptions to this general requirement.

CONDUCTOR SIZING AND OVERCURRENT PROTECTION *(cont.)*

Code	Description
E3705.4.4	Conductors in NM cable must be rated at 90°C. Types NM, NMC, and NMS marked as NM-B, NMC-B, and NMS-B have this rating.

TABLE E3705.5.3 OVERCURRENT-PROTECTION RATING

COPPER		ALUMINUM OR COPPER-CLAD ALUMINUM	
Size (AWG)	Maximum overcurrent-protection-device rating[a] (amps)	Size (AWG)	Maximum overcurrent-protection-device rating[a] (amps)
14	15	12	15
12	20	10	25
10	30	8	30

a. The maximum overcurrent-protection-device rating shall not exceed the conductor allowable ampacity determined by the application of the correction and adjustment factors in accordance with Sections E3705.2 and E3705.3.

© 2018 IRC®, International Code Council®

ELECTRICAL

OVERCURRENT DEVICES

Common Cable Types and Sizes for Various Circuit Breakers

	15 amp	20 amp	30 amp	40 amp	50 amp
Plugs and switches in habitable areas	NM $14/2$ with ground copper	NM $12/2$ with ground copper	NP	NP	NP
Plugs and switches in kitchens, laundry, bathrooms, and dining rooms	NP	NM $12/2$ with ground copper	NP	NP	NP
Microwave	NP	NM $12/2$ with ground copper	NP	NP	NP
Dishwasher and disposer	NM $14/2$ with ground copper	NM $12/2$ with ground copper	NP	NP	NP
Refrigerator and freezer	NM $14/2$ with ground copper	NM $12/2$ with ground copper	NP	NP	NP
Oven and range	NP	NP	NM $10/3$ with ground copper	NM $8/3$ with ground copper	NM $6/3$ with ground copper
Free-standing range/oven	NP	NP	NP	NM $8/3$ with ground copper	NM $6/3$ with ground copper
Water heater	NP	NP	NM $10/3$ with ground copper	NP	NP
Clothes dryer	NP	NP	NM $10/3$ with ground copper	NP	NP
Hydro-massage tub	NP	NM $12/2$ with ground copper	NP	NP	NP

NP – Not Permitted

OVERCURRENT DEVICES *(cont.)*

Selection of Overcurrent Devices

1	Total Volts	Connected Watts	Determine the wattage and voltage demand of the appliance. This information is on the nameplate of the device.
	120	4000	
2	Determine the current (amperage) demand of the appliance either by the nameplate rating or by dividing the watts by the voltage $^{4000}/_{120} = 33.33$ amps.		
3	Select the next size up for overcurrent protection. In this case, 40 amps.		

Circuit Size	Copper NM Cable	Aluminum or Copper Clad
15 amps	$^{14}/_2$ with ground	$^{12}/_2$ with ground
20 amps	$^{12}/_2$ with ground	$^{10}/_2$ with ground
25 amps	$^{10}/_3$ with ground	$^{10}/_2$ with ground
30 amps	$^{10}/_3$ with ground	$^{8}/_2$ with ground
40 amps	$^{8}/_3$ with ground	$^{6}/_3$ with ground

ELECTRICAL

ABBREVIATIONS

AC	Armored cable	NM	Nonmetallic sheathed cable
EMT	Electrical metallic tubing	RNC	Rigid nonmetallic conduit
ENT	Electrical nonmetallic tubing	RMC	Rigid metallic conduit
FMC	Flexible metal conduit	SE	Service entrance cable
IMC	Intermediate metal conduit	SR	Surface raceways
LFC	Liquidtight flexible conduit	UF	Underground feeder cable
MC	Metal-clad cable	USE	Underground service cable

Code	Description
E3801.3	All conductors of a circuit, including equipment grounding conductors and bonding conductors, shall be contained in the same raceway, trench, cable, or cord.

TABLE E3801.4 ALLOWABLE APPLICATIONS FOR WIRING METHODS[a, b, c, d, e, f, g, h, i, j, k]

ALLOWABLE APPLICATIONS (application allowed where marked with an "A")	AC	EMT	ENT	FMC	IMC RMC RNC RTRC	LFC[a, g]	MC	NM	SR	SE	UF	USE
Services	—	A	A[h]	A[i]	A	A[i]	A	—	—	A	—	A
Feeders	A	A	A	A	A	A	A	A	—	A[b]	A	A[b]
Branch circuits	A	A	A	A	A	A	A	A	A	A[c]	A	—
Inside a building	A	A	A	A	A	A	A	A	A	A	A	—
Wet locations exposed to sunlight	—	A	A[h]	—	A	A	A	—	—	A	A[e]	A[e]
Damp locations	—	A	A[d]	A	A	A	A	—	—	A	A	A
Embedded in noncinder concrete in dry location	—	A	A	A	A	A[i]	—	—	—	—	—	—
In noncinder concrete in contact with grade	—	A[f]	A	—	A[f]	A[i]	—	—	—	—	—	—
Embedded in plaster not exposed to dampness	A	A	A	A	A	A	A	—	—	A	A	—
Embedded in masonry	—	A	A	—	A[f]	A	A	—	—	—	—	—
In masonry voids and cells exposed to dampness or below grade line	—	A[f]	A	A[d]	A[f]	A	A	—	—	A	A	—
Fished in masonry voids	A	—	—	A	—	A	A	A	—	A	A	—
In masonry voids and cells not exposed to dampness	A	A	A	A	A	A	A	A	—	A	A	—
Run exposed	A	A	A	A	A	A	A	A	A	A	A	—
Run exposed and subject to physical damage	—	—	—	—	A[g]	—	—	—	—	—	—	—
For direct burial	—	A[f]	—	—	A[f]	A	A[f]	—	—	—	A	A

For SI: 1 foot = 304.8 mm.

a. Liquid-tight flexible nonmetallic conduit without integral reinforcement within the conduit wall shall not exceed 6 feet in length.
b. Type USE cable shall not be used inside buildings.
c. The grounded conductor shall be insulated.
d. Conductors shall be a type approved for wet locations and the installation shall prevent water from entering other raceways.
e. Shall be listed as "Sunlight Resistant."
f. Metal raceways shall be protected from corrosion and approved for the application. Aluminum RMC requires approved supplementary corrosion protection.
g. RNC shall be Schedule 80. RTRC shall be RTRC-XW.
h. Shall be listed as "Sunlight Resistant" where exposed to the direct rays of the sun.
i. Conduit shall not exceed 6 feet in length.
j. Liquid-tight flexible nonmetallic conduit is permitted to be encased in concrete where listed for direct burial and only straight connectors listed for use with LFNC are used.
k. In wet locations under any of the following conditions:
  1. The metallic covering is impervious to moisture.
  2. A lead sheath or moisture-impervious jacket is provided under the metal covering.
  3. The insulated conductors under the metallic covering are listed for use in wet locations and a corrosion-resistant jacket is provided over the metallic sheath.

© 2018 IRC®, International Code Council®

Code	Description
E3801.4	Wiring methods must comply with Table E3801.4.

ABOVEGROUND INSTALLATIONS

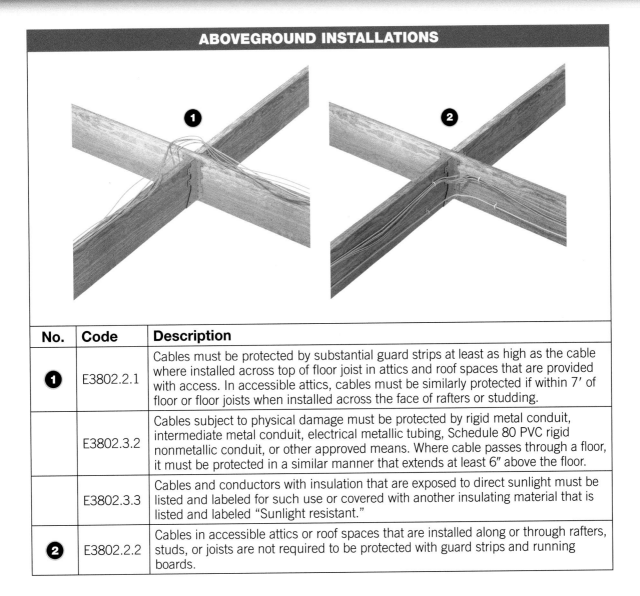

No.	Code	Description
1	E3802.2.1	Cables must be protected by substantial guard strips at least as high as the cable where installed across top of floor joist in attics and roof spaces that are provided with access. In accessible attics, cables must be similarly protected if within 7′ of floor or floor joists when installed across the face of rafters or studding.
	E3802.3.2	Cables subject to physical damage must be protected by rigid metal conduit, intermediate metal conduit, electrical metallic tubing, Schedule 80 PVC rigid nonmetallic conduit, or other approved means. Where cable passes through a floor, it must be protected in a similar manner that extends at least 6″ above the floor.
	E3802.3.3	Cables and conductors with insulation that are exposed to direct sunlight must be listed and labeled for such use or covered with another insulating material that is listed and labeled "Sunlight resistant."
2	E3802.2.2	Cables in accessible attics or roof spaces that are installed along or through rafters, studs, or joists are not required to be protected with guard strips and running boards.

ELECTRICAL

ABOVEGROUND INSTALLATIONS

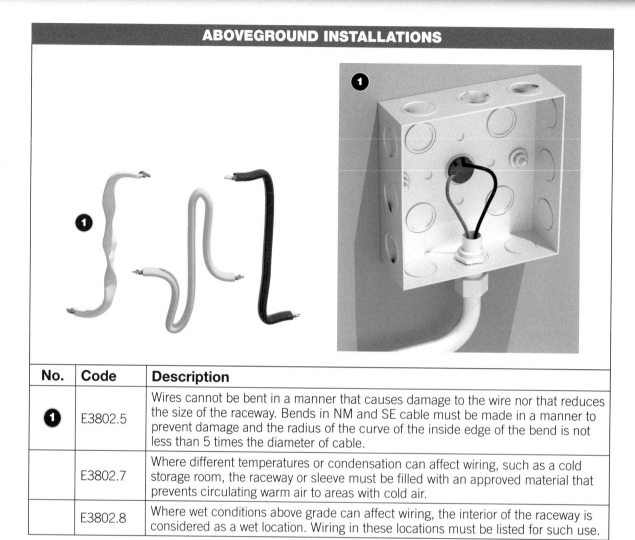

No.	Code	Description
1	E3802.5	Wires cannot be bent in a manner that causes damage to the wire nor that reduces the size of the raceway. Bends in NM and SE cable must be made in a manner to prevent damage and the radius of the curve of the inside edge of the bend is not less than 5 times the diameter of cable.
	E3802.7	Where different temperatures or condensation can affect wiring, such as a cold storage room, the raceway or sleeve must be filled with an approved material that prevents circulating warm air to areas with cold air.
	E3802.8	Where wet conditions above grade can affect wiring, the interior of the raceway is considered as a wet location. Wiring in these locations must be listed for such use.

UNDERGROUND INSTALLATIONS

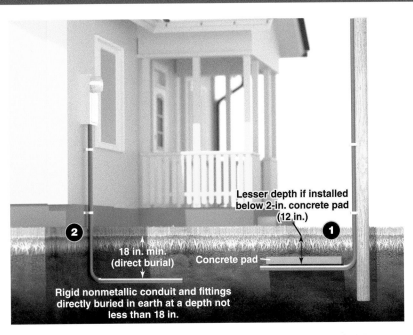

Lesser depth if installed below 2-in. concrete pad (12 in.)

18 in. min. (direct burial)

Concrete pad

Rigid nonmetallic conduit and fittings directly buried in earth at a depth not less than 18 in.

No.	Code	Description
❶	E3803.1	Cable or raceways that are buried must have adequate cover. Table E3803.1 in the IRC code book sets out minimum cover for different conditions and type of wiring.
❷	E3803.2	Service conductors that are buried underground more than 18″ and not encased in concrete must have a warning ribbon not less than 12″ above the installation, identifying their presence.
	E3803.3	Conductors and cables emerging from the ground must be protected from damage by enclosures or raceways that extend from a depth required for cover to a point at least 8′ above finished grade but need not exceed 18″ below finished grade.
	E3803.4	Cables of conductors direct buried below grade may be spliced or tapped without splice boxes.
	E3803.5	Backfill over cables or conductors must not contain large rock, paving material, cinders, large, or angular substances or corrosive materials. Backfill must be able to be properly compacted.

You Should Know

- The interior of enclosures or raceways installed underground shall be considered to be a wet location. Insulated conductors and cables installed in such enclosures or raceways in underground installations shall be listed for use in wet locations; Section E3803.10 in the IRC code book.

ELECTRICAL

RECEPTACLE OUTLETS

Wall receptacles serve the spaces for 6 ft on each side of the receptacle. Therefore, the maximum spacing between wall receptacles is 12 ft.

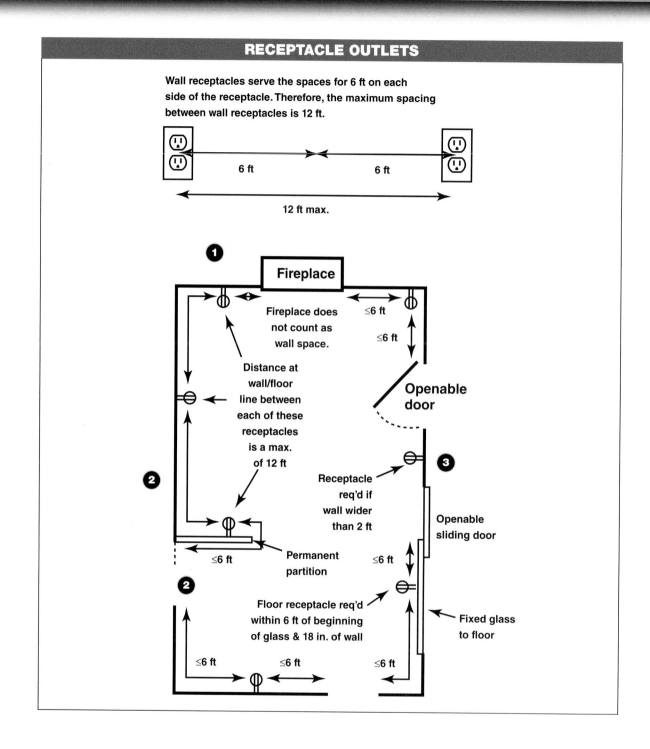

Fireplace

Fireplace does not count as wall space.

≤6 ft
≤6 ft

Openable door

Distance at wall/floor line between each of these receptacles is a max. of 12 ft

Receptacle req'd if wall wider than 2 ft

Openable sliding door

≤6 ft

Permanent partition

≤6 ft

Floor receptacle req'd within 6 ft of beginning of glass & 18 in. of wall

Fixed glass to floor

≤6 ft   ≤6 ft   ≤6 ft

RECEPTACLE OUTLETS *(cont.)*

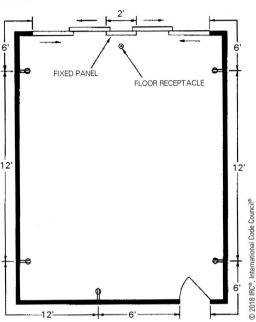

Figure E3901.2
General Use Receptacle Distribution
For SI: 1 foot = 304.8 mm

© 2018 IRC®, International Code Council®

No.	Code	Description
❶	E3901.2	Electrical receptacles must be installed along a wall space in every kitchen, family room, dining room, living room, parlor, library, den, sunroom, bedroom, recreation room, or similar rooms (see Figure E3901.2 above).
❷	E3901.2.1	Receptacles must be spaced along a wall to ensure that no floor line or wall space is more than 6′ from a receptacle.
❸	E3901.2.2	A wall space includes: 1. Any space that is 2′ or more in width including space around corners including wall spaces created by doors, around corners, fireplaces, fixed cabinets that do not have countertop, or similar work surfaces. 2. Space occupied by fixed panels in exterior walls (not sliding panels). 3. Space formed by fixed room dividers, such as railing or bar-type counters.
	E3901.2.3	Floor receptacles cannot be used as required outlets unless they are within 18″ of the wall.

You Should Know

- Receptacle outlets required by this section shall be in addition to any receptacle that is:
  1. Part of a luminaire or appliance;
  2. Located within cabinets or cupboards;
  3. Controlled by a wall switch in accordance with Section E3903.2 in the IRC code book, exception 1; or
  4. Located over 5.5′ above the floor.

ELECTRICAL

OUTLETS: 120-VOLT RECEPTACLE
Duplex Receptacle

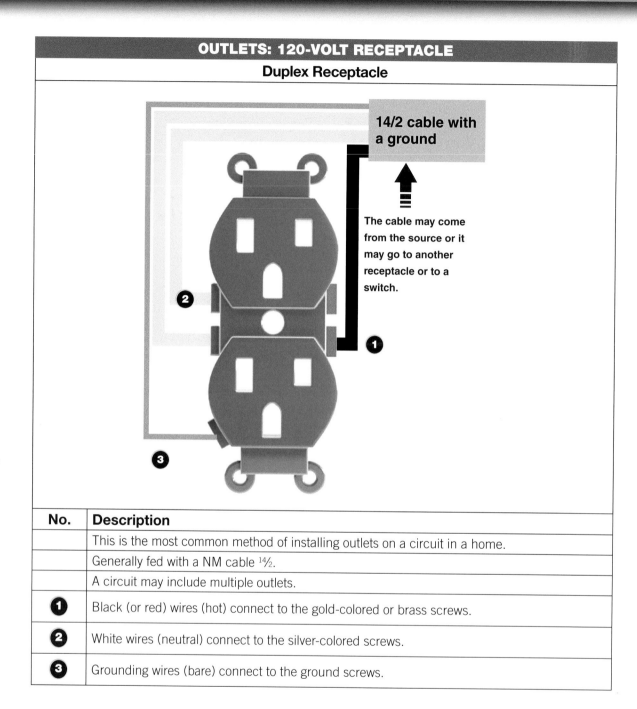

14/2 cable with a ground

The cable may come from the source or it may go to another receptacle or to a switch.

No.	Description
	This is the most common method of installing outlets on a circuit in a home.
	Generally fed with a NM cable $^{14}/_2$.
	A circuit may include multiple outlets.
❶	Black (or red) wires (hot) connect to the gold-colored or brass screws.
❷	White wires (neutral) connect to the silver-colored screws.
❸	Grounding wires (bare) connect to the ground screws.

APPLIANCE CIRCUIT RECEPTACLE OUTLETS
(KITCHEN AND DINING ROOM)

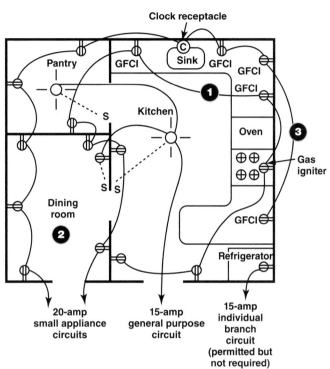

Clock receptacle

Pantry

GFCI

Sink

GFCI

GFCI

GFCI

Kitchen

Oven

Gas igniter

Dining room

GFCI

Refrigerator

20-amp small appliance circuits

15-amp general purpose circuit

15-amp individual branch circuit (permitted but not required)

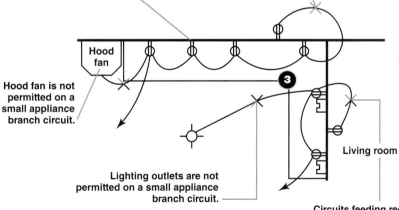

Dedicated clock outlet is permitted on a small appliance branch circuit.

Outdoor receptacles not allowed on small-appliance branch circuits.

Hood fan

Hood fan is not permitted on a small appliance branch circuit.

Living room

Lighting outlets are not permitted on a small appliance branch circuit.

Circuits feeding receptacles in kitchens, pantries, breakfast rooms, dining rooms, or similar areas shall not feed receptacles outside of those areas.

See E3901, or refer to the NEC as permitted by Chapter 34 of the IRC.

ELECTRICAL

No.	Code	Description
1	E3901.3	Two or more 20-ampere appliance branch circuits are needed in the kitchen, pantry, breakfast room, dining room, or similar areas including refrigeration appliances.
2	E3901.3.1	With some exceptions, the two or more required appliance branch circuits may not serve any other outlets.
3	E3901.3.2	At least two appliance branch circuits must be provided that serve receptacles in kitchens that serve countertop surfaces.
		Additional appliance branch circuits may be provided for in kitchen, pantry, breakfast room, dining room, or similar areas.

APPLIANCE CIRCUIT RECEPTACLE OUTLETS (KITCHEN AND DINING ROOM) *(cont.)*

You Should Know

- Countertop spaces separated by range tops, refrigerators, or sinks shall be considered as separate countertop spaces.
- There are exceptions that allow a refrigerator and other specific appliances to be served with separate single 15-ampere branch circuits.

KITCHEN COUNTERTOP RECEPTACLES

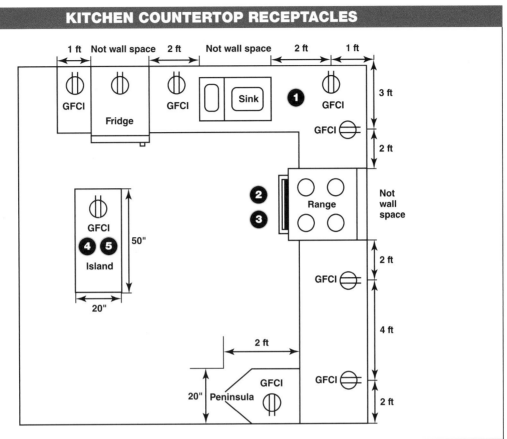

No.	Code	Description
❶		Spacing of receptacles around countertops in kitchen, pantry, breakfast, dining room, or similar areas must be according to Figures E3901.4.1 through E3901.4.5.
	E3901.4.4	Two separate countertop spaces are created when in a peninsula or island, a range, cooking unit, or sink causes the width of the countertop to be less than 12".
❷		Each separate countertop space must comply with the required spacing.
❸	E3901.4.1	A receptacle must be installed along each wall countertop space 12" or wider and spaced so that no point along wall line is more than 24" from a receptacle or space. The exception to this general rule is spaces behind range or sink (see Figure E3901.4.1).
❹	E3901.4.2	At least one receptacle outlet is needed in an island countertop that is 24" or longer with a short dimension of 12" or greater.
❺	E3901.4.3	At least one receptacle outlet is needed in a peninsula countertop that is 24" or longer with a short dimension of 12" or greater.

KITCHEN COUNTERTOP RECEPTACLES *(cont.)*

Figure E3901.4.1
Determination of Area Behind Sink or Range
For SI: 1 inch = 25.4 mm

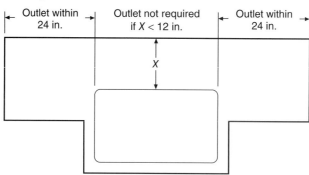

Sink, range, or counter-mounted cooking unit extending from face of counter

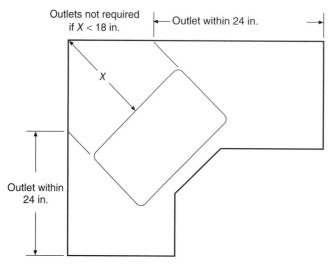

Sink, range, or counter-mounted cooking unit mounted in corner

© 2018 IRC®, International Code Council®

Code	Description
E3901.4.5	Receptacle outlets cannot be more than 20″ higher than the surface of the countertop or work surface.
	Receptacle outlets in work surfaces or countertops must be listed for such use.
E3901.5	Receptacle outlets dedicated to an appliance must be within 6′ of the appliance, such as a washing machine.

You Should Know

- Receptacle outlets rendered not readily accessible by appliances fastened in place, appliance garages, sinks, or rangetops as addressed in the exception to Section E3901.4.1 in the IRC code book, or appliances occupying dedicated space shall not be considered as these required outlets.

OTHER RECEPTACLE OUTLETS (BATHROOM, OUTSIDE, AND BASEMENTS)

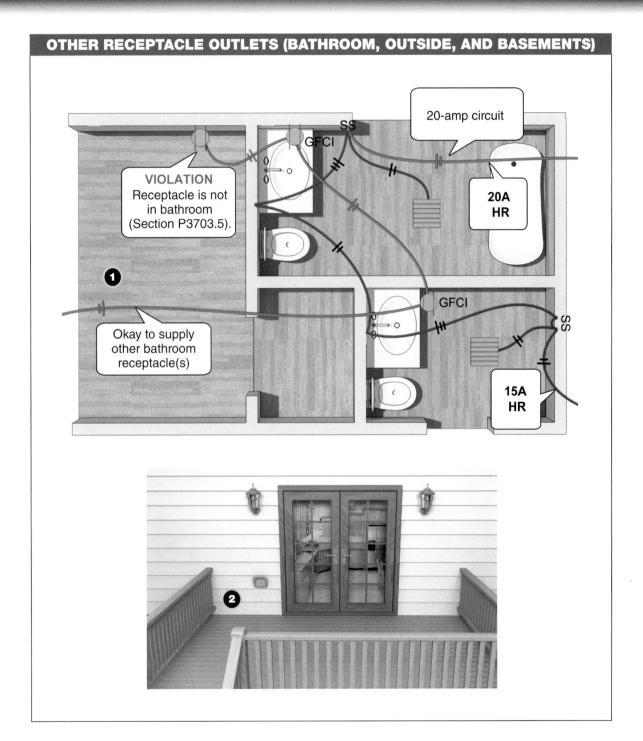

OTHER RECEPTACLE OUTLETS (BATHROOM, OUTSIDE, AND BASEMENTS) *(cont.)*

No.	Code	Description
1	E3901.6	Bathrooms must have at least one receptacle outlet within 36" of the outside edge of each lavatory basin. This receptacle outlet must be installed on the wall or partition adjacent to the basin, on the countertop, or on the side or face of the cabinet not more than 12" below the top of the basin or basin countertop.
		Bathroom receptacle outlets in countertops must be listed for the installation.
2	E3901.7	Receptacles are required outside in the front and rear of the building. The receptacles must be accessible while standing at grade level.
		A receptacle outlet must be provided within the perimeter of balconies, decks, or porches that are accessible from within the building. Receptacles cannot be higher than 6'6" above the floor surface of the balcony, deck, or porch.

You Should Know

- Outlets in these areas generally must be protected by a ground-fault circuit interrupter.

OTHER RECEPTACLE OUTLETS

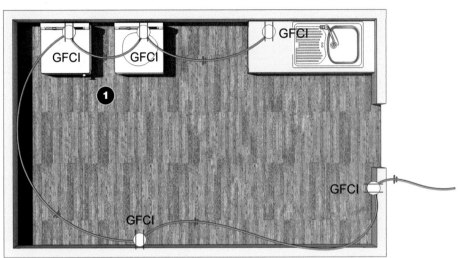

No.	Code	Description
①	E3901.8	Areas for laundry appliances must have at least one receptacle outlet.
	E3901.9	Not less than one receptacle outlet, in addition to any provided for specific equipment, shall be installed in each separate unfinished portion of a basement, in each vehicle bay not more than 5.5′ above the floor in an attached garage, in each vehicle bay of a detached garage that is provided with electrical power, and accessory buildings provided with electrical power.
	E3901.10	Hallways more than 10′ long must have at least one receptacle outlet.
	E3901.11	Foyers greater than 60 ft², not part of hallways, must have a receptacle outlet for each unbroken wall section 3′ or more in width.
	E3901.12	HVAC equipment must be provided with a receptacle outlet for service installed in an accessible location. It must be on the same level and within 25′ of the equipment. It cannot be connected to the load side of the equipment.

You Should Know

- A receptacle outlet shall not be required for the servicing of evaporative coolers.
- All receptacles within laundry area must have GFCI protection for personnel.

ELECTRICAL

GROUND-FAULT CIRCUIT INTERRUPTER

Garage/Workshop

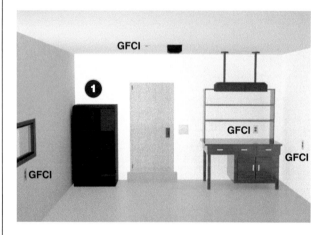

No.	Code	Description
	E3902.1	All 15- or 20-ampere receptacles in a bathroom must have GFCI protection for personnel.
1	E3902.2	All 15- or 20-ampere receptacles in a garage or grade portions of unfinished accessory buildings for storage or work area must have GFCI protection for personnel.
	E3902.3	All 15- or 20-ampere receptacles outdoors must have GFCI protection for personnel.
2	E3902.4	All 15- or 20-ampere receptacles in a crawl space at or below grade must have GFCI protection for personnel. Lighting outlets not exceeding 120 volts must have GFCI protection.

GROUND-FAULT CIRCUIT INTERRUPTER *(cont.)*

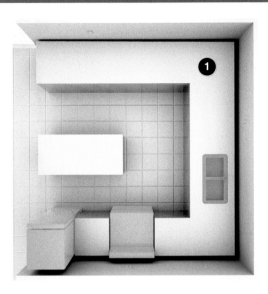

No.	Code	Description
	E3902.5	All 15- or 20-ampere receptacles in an unfinished basement must have GFCI protection for personnel.
❶	E3902.6	All 15- or 20-ampere receptacles that serve countertop surfaces in kitchens must have GFCI protection for personnel.
	E3902.7	All 15- or 20-ampere receptacles within 6' of the top, inside edge of the bowl of a sink must have GFCI protection for personnel.

ELECTRICAL

GROUND-FAULT CIRCUIT INTERRUPTER *(cont.)*

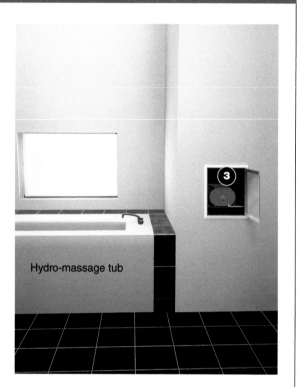

Hydro-massage tub

No.	Code	Description
	E3902.9	Receptacles in laundry area must have GFCI protection for personnel.
	E3902.10	GFCI protection is required for outlets that supply dishwashers.
❶	E3902.11	All 15- or 20-ampere receptacles installed in boathouses must have GFCI protection for personnel.
❷	E3902.12	Outlet receptacles that supply boat hoists must have GFCI protection for personnel.
❸	E3902.13	Electrically heated cables embedded in concrete or poured masonry floors in bathrooms, kitchens, hydro-massage bathtubs, spa, and hot tub locations must be protected by GFCI type protection for personnel.
	E3902.14	All GFCI devices must be in a readily accessible location.

OUTLETS DUPLEX: 120-VOLT GFCI RECEPTACLE

Duplex GFCI Receptacle

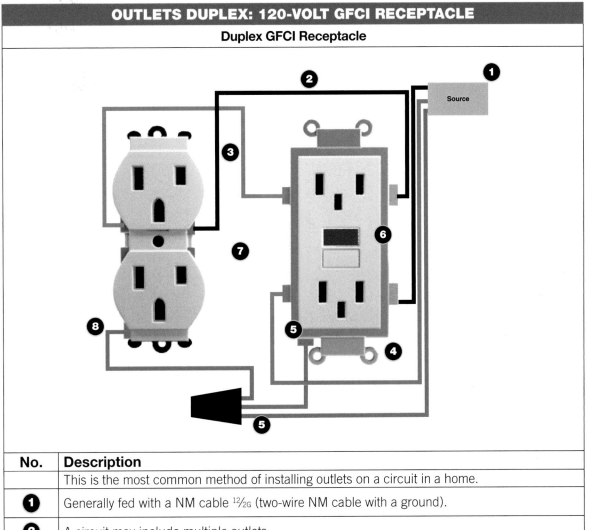

No.	Description
	This is the most common method of installing outlets on a circuit in a home.
❶	Generally fed with a NM cable $^{12}/_{2G}$ (two-wire NM cable with a ground).
❷	A circuit may include multiple outlets.
❸	Black wires (hot) connect to the gold-colored or brass screws.
❹	White wires (neutral) connect to the silver-colored screws.
❺	Grounding wires (bare) connect to the ground screws.
❻	Black wire from the power supply will be connected to the gold screw marked line.
❼	White wire from the power supply will be connected to the silver screw marked line.
❽	Bare wire from the power supply will be connected to the screw marked ground.
	For multiple outlets on the same circuits, the hot and neutral wires (black and white) are connected to load terminal screws that lead to another receptacle. Ground wire feeds through.

ELECTRICAL

ARC-FAULT CIRCUIT INTERRUPTER

No.	Code	Description
❶	E3902.16	Branch circuits that supply 15- and 20-ampere outlets installed in kitchens, family rooms, dining rooms, living rooms, parlors, libraries, dens, bedrooms, sunrooms, recreations rooms, closets, hallways, laundry areas and similar rooms or areas shall be protected by any of the following: 1. A listed combination type arc fault circuit interrupter installed to provide protection of the entire branch circuit. 2. A listed combination type AFCI installed at the origin of the branch circuit in combination with a listed outlet branch circuit type arc fault circuit interrupter installed at the first outlet. 3. A listed supplemental arc protection circuit breaker installed at the origin of the branch circuit in combination with a listed outlet branch circuit type arc fault circuit interrupter installed at the first outlet with three conditions (see code reference). 4. A listed outlet branch circuit type arc fault circuit interrupter installed at the first outlet of the branch circuit in combination with a listed branch circuit overcurrent protective device where four conditions are met (see code reference).

You Should Know

- The ground-fault circuit interrupter, or GFCI, and arc-fault circuit interrupter, or AFCI, are two different electrical devices that perform different types of protection. A GFCI is designed to prevent electrical shock and is typically in damp and outdoor locations where a ground fault could occur. The AFCI is designed to prevent fires caused by arcing.

LIGHTING OUTLETS

ELECTRICAL

No.	Code	Description
	E3903.2	At least one wall switch-controlled lighting outlet must be in every habitable room, kitchen, and bathroom, except: 1. Receptacles controlled by a wall switch (not in kitchens and bathrooms), or 2. Where controlled by occupancy sensors that are in addition to wall switches or located in a customary wall switch location.
❶		At least one wall switch-controlled lighting outlet is needed in hallways, stairways, attached garages, and detached garages with power.
	E3903.3	At least one wall-switch-controlled lighting outlet is necessary to provide lighting on the exterior side of every outside egress door with grade access. This includes attached garages and detached garages with power.
		Lighting, in interior stairs with six or more risers, must be controlled by a wall switch at each floor level. This includes entryways.
❷	E3903.4	At least one lighting outlet must be installed in attics, crawl spaces, utility rooms, and basements used for storage or contain equipment. The lighting outlet must be controlled by a wall switch or an integral switch. The switch should be near the entry to these spaces. The lighting should be at or near the equipment requiring repairs. Dimmer switches, generally, are not permitted for these lighting controls.

LIGHTING OUTLETS *(cont.)*

You Should Know

- In hallways, stairways, and at outdoor egress doors, remote, central, or automatic control of lighting shall be permitted.

GENERAL INSTALLATION REQUIREMENTS

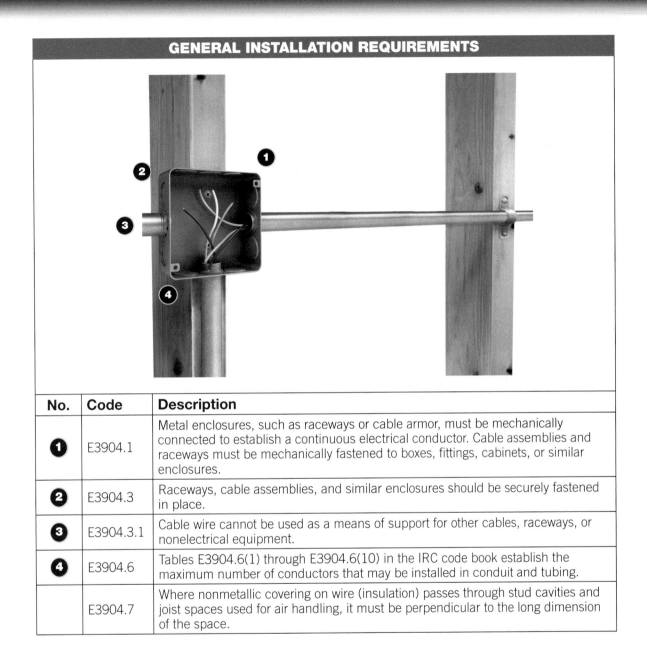

No.	Code	Description
1	E3904.1	Metal enclosures, such as raceways or cable armor, must be mechanically connected to establish a continuous electrical conductor. Cable assemblies and raceways must be mechanically fastened to boxes, fittings, cabinets, or similar enclosures.
2	E3904.3	Raceways, cable assemblies, and similar enclosures should be securely fastened in place.
3	E3904.3.1	Cable wire cannot be used as a means of support for other cables, raceways, or nonelectrical equipment.
4	E3904.6	Tables E3904.6(1) through E3904.6(10) in the IRC code book establish the maximum number of conductors that may be installed in conduit and tubing.
	E3904.7	Where nonmetallic covering on wire (insulation) passes through stud cavities and joist spaces used for air handling, it must be perpendicular to the long dimension of the space.

ELECTRICAL

BOXES, CONDUIT BODIES, AND FITTINGS

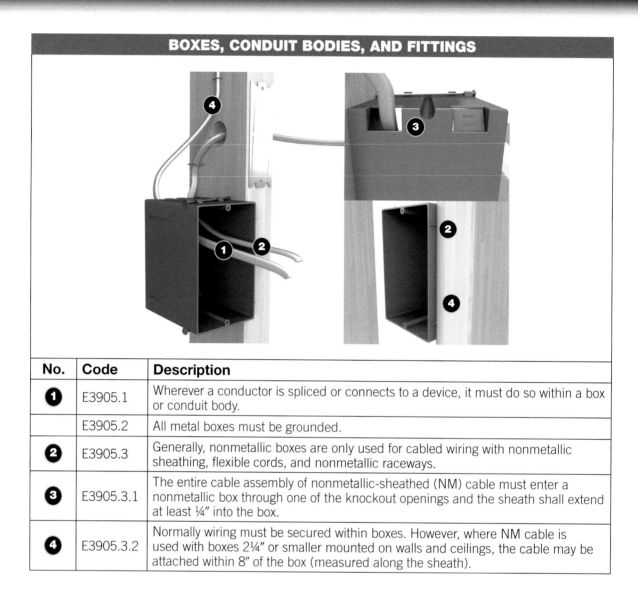

No.	Code	Description
❶	E3905.1	Wherever a conductor is spliced or connects to a device, it must do so within a box or conduit body.
	E3905.2	All metal boxes must be grounded.
❷	E3905.3	Generally, nonmetallic boxes are only used for cabled wiring with nonmetallic sheathing, flexible cords, and nonmetallic raceways.
❸	E3905.3.1	The entire cable assembly of nonmetallic-sheathed (NM) cable must enter a nonmetallic box through one of the knockout openings and the sheath shall extend at least ¼" into the box.
❹	E3905.3.2	Normally wiring must be secured within boxes. However, where NM cable is used with boxes 2¼" or smaller mounted on walls and ceilings, the cable may be attached within 8" of the box (measured along the sheath).

You Should Know

- Fittings and connectors shall be used only with the specific wiring methods for which they are designed and listed.
- While the code does not prescriptively limit the number of outlet devices on a circuit, normally 8–10 devices (outlets and lights) are installed on each branch circuit. Devices such as higher-wattage lights should be on an individual circuit.

BOXES, CONDUIT BODIES, AND FITTINGS

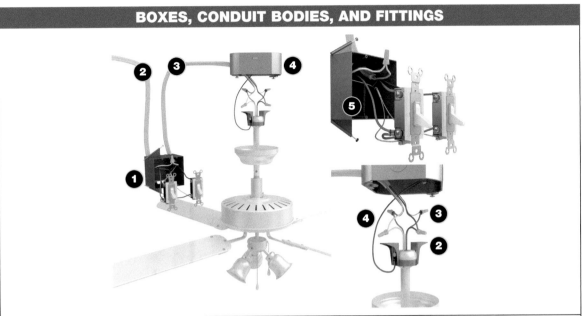

No.	Code	Description
1	E3905.4	All outlet and device boxes shall have an approved depth to allow equipment installed within them to be mounted properly and without the likelihood of damage to conductors within the box.
2	E3905.6.1	Boxes used at luminaire outlet or lampholder outlets in or on a vertical surface shall be identified and marked on the interior of the box to indicate the maximum weight that is permitted to be supported by the box if other than 50 pounds. Boxes used to support luminaire or lamps in a ceiling must be capable of supporting the intended load if other than 50 pounds and must be marked for that purpose.
3	E3905.6.2	Every outlet used exclusively for lighting must be able to support a lamp weighing 50 pounds and designed so that the luminaire can be attached unless the box is listed for not less than the weight supported.
4	E3905.8	Outlet boxes used only for supporting paddle type ceiling fans must be marked for that purpose and not support ceiling suspended paddle fans that weigh more than 70 pounds. Any box designed to support these fans that weigh more than 35 pounds must be marked with the weight of the fan to be supported.
5	E3906.6	Openings in wall or ceiling surfaces where boxes are installed that have a flush-type cover must not have gaps or open spaces larger than ⅛″ around the box's edge. Any metal covers and plates must be grounded.

You Should Know

- Any outlet box must be listed for the use when supporting ceiling fans.

ELECTRICAL

BOX FILL

TABLE E3905.12.2.1 VOLUME ALLOWANCE REQUIRED PER CONDUCTOR

SIZE OF CONDUCTOR	FREE SPACE WITHIN BOX FOR EACH CONDUCTOR (cubic inches)
18 AWG	1.50
16 AWG	1.75
14 AWG	2.00
12 AWG	2.25
10 AWG	2.50
8 AWG	3.00
6 AWG	5.00

For SI: 1 cubic inch = 16.4 cm^3.

© 2018 IRC®, International Code Council®

TYPICAL CALCULATIONS

Item	Quantity		Volume		Total
Hots	_____	×	_____	=	_____
Neutrals	_____	×	_____	=	_____
Grounds	_____	×	_____	=	_____
Switch or Receptacle	_____	×	_____	=	_____

Total Volume Requirements for This Box = _____

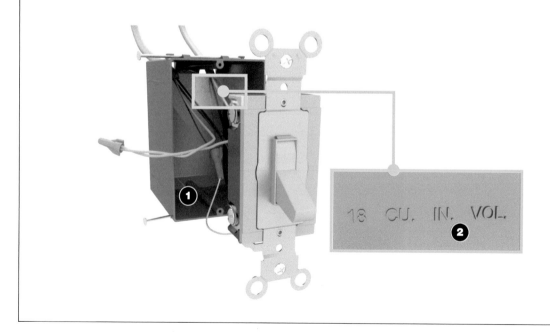

BOX FILL *(cont.)*		
No.	**Code**	**Description**
❶	E3905.12.1.1	Boxes are limited to a maximum number of wires or devices based on their size. Table E3905.12.2.1 depicts that limit for standard size boxes.
❷	E3905.12.2.1	Some rules apply to box fill calculation: 1. Each conductor from outside the box and terminates or is spliced counts as one. 2. Each loop or coil of unbroken conductor longer than 6″ (Section E3406.11.3) counts as two. 3. A conductor that does not leave the box need not be counted. 4. Each conductor that passes through the box without a splice counts as one.

You Should Know

- Count the number of wires (hot, neutral, and ground) and enter the quantity and volume for each.
- Count the number of devices such as outlets or switches and enter the quantity and volume for each.
- Compare results to the maximum capacity of the specific box in Table 3905.12.1.

ELECTRICAL

CABINETS AND PANELBOARDS

TABLE E3907.8 PERCENT OF CROSS SECTION OF CONDUIT AND TUBING FOR CONDUCTORS	
NUMBER OF CONDUCTORS	**MAXIMUM PERCENT OF CONDUIT AND TUBING AREA FILLED BY CONDUCTORS**
1	53
2	31
Over 2	40

© 2018 IRC®, International Code Council®

Code	Description
E3907.8	Nonmetallic sheathed cables are allowed to enter the top of a surface mounted enclosure inside a rigid raceway. The raceway may be one or more sections and at least 18" long but not more than 10' in overall length if all of the following conditions are met:
E3907.8 exceptions 1–7.	1. Each cable is fastened within 12" of the outer end of the raceway. 2. The raceway extends directly above the enclosure while not passing through a structural ceiling. 3. An accessible fitting is installed on each end of the raceway to protect the cable from abrasion. 4. The raceway is sealed or plugged at the outer end so that there is no access to the raceway. 5. The cable sheathing is continuous through the raceway and into the enclosure at least ¼". 6. The raceway is secured at the outer end and other locations as required. 7. The allowable cable fill cannot be more than allowed by Table E3907.8 and shall be considered as a complete conduit or tubing system. A multiconductor cable having two or more conductors is regarded as a single conductor for calculating conduit fill area. For cables that have elliptical cross sections, the cross-sectional area calculation shall be based on the major diameter of the ellipse as a circle diameter.

GROUNDING

TABLE E3908.12 EQUIPMENT GROUNDING CONDUCTOR SIZING

RATING OR SETTING OF AUTOMATIC OVERCURRENT DEVICE IN CIRCUIT AHEAD OF EQUIPMENT, CONDUIT, ETC., NOT EXCEEDING THE FOLLOWING RATINGS (amperes)	MINIMUM SIZE	
	Copper wire No. (AWG)	Aluminum or copper-clad aluminum wire No. (AWG)
15	14	12
20	12	10
60	10	8
100	8	6
200	6	4
300	4	2
400	3	1

© 2018 IRC®, International Code Council®

No.	Code	Description
1	E3908.8	Equipment grounding conductor that is with or enclosing circuit conductors must be one or more of the following: 1. A copper, aluminum, or copper-clad conductor. This conductor is solid or stranded; insulated, covered, or bare; and in the form of a wire or a bus bar of any shape. 2. Rigid metal conduit. 3. Intermediate metal conduit. 4. Electrical metallic tubing. 5. Armor of Type AC cable in accordance with Section E3908.4 in the IRC code book.

ELECTRICAL

		GROUNDING *(cont.)*
No.	**Code**	**Description**

	E3908.8	6. Type MC cable that provides an effective ground-fault current path with one of the following:
		   6.1. It contains an insulated or uninsulated equipment grounding conductor in compliance with item 1 of this section.
		   6.2. The combined metallic sheath and uninsulated equipment grounding/bonding conductor of interlocked metal tape Type MC cable that is listed and identified as an equipment grounding conductor.
		   6.3. The metallic sheath of the combined metallic sheath and equipment grounding conductors of the smooth or corrugated tube-type MC that is listed and identified as an equipment grounding conductor.
		7. Other electrically continuous metal raceways and auxiliary gutters.
		8. Surface metal raceways listed for grounding.

		Copper, aluminum, and copper-clad aluminum wire type equipment grounding conductors must be designed using Table E3908.12, but they do not need to be larger than the circuit conductors supplying the equipment.

You Should Know

- The grounding equipment conductor (ECG) is the conductive path installed to connect normally non current-carrying metal parts of equipment together and to the system grounded conductor, the grounding electrode conductor, or both.

SWITCHES

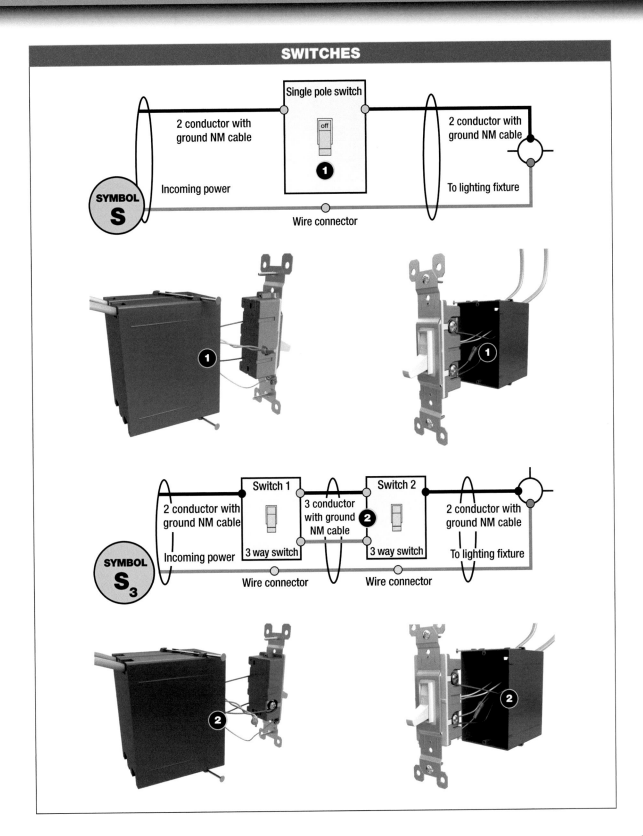

Single pole switch

off

2 conductor with ground NM cable

2 conductor with ground NM cable

SYMBOL S

Incoming power

To lighting fixture

Wire connector

Switch 1

Switch 2

2 conductor with ground NM cable

3 conductor with ground NM cable

2 conductor with ground NM cable

SYMBOL S₃

Incoming power

3 way switch

3 way switch

To lighting fixture

Wire connector

Wire connector

ELECTRICAL

SWITCHES (cont.)

No.	Code	Description
1	E4001.3	Generally, switches must indicate whether the circuit is open or closed. This applies to general use, motor circuits, and circuit breakers. If the switch is installed vertically, the up position is the ON position.
2	E4001.9	Methods for wiring three- and four-way switches must be so that switching occurs only for the ungrounded circuit conductor.
	E4001.11.1	Generally, metal faceplates must be grounded. Snap switches (including dimmers) must be connected to an equipment grounding conductor. In addition, metal faceplates must be connected to the equipment grounding conductor and grounded. Snap switches are considered effectively grounded if any one of the following conditions is satisfied: 1. The switch is mounted with metal screws to a metal box or metal cover connected to an equipment grounding conductor or to a nonmetallic box with integral means for connecting to an equipment grounding conductor. 2. An equipment grounding conductor or bonding jumper is connected to an equipment grounding termination of the snap switch. **Exception:** 1. Where a means to connect to an equipment grounding conductor does not exist within the snap-switch enclosure or where the wiring method does not include or provide an equipment grounding conductor, a snap switch without a grounding connection to an equipment grounding conductor shall be permitted for replacement purposes only. A snap switch wired under the provisions of this exception and located within 8′ vertically or 5′ horizontally or a ground or exposed grounded metal objects shall be provided with a faceplate of nonconducting, noncombustible material with a nonmetallic attachment screw, unless the switch mounting strap or yoke is nonmetallic or the circuit is protected by a ground-fault circuit interrupter. 2. Listed kits or listed assemblies shall not be required to be connected to an equipment grounding conductor if all of the following conditions apply: 2.1. The device is provided with a nonmetallic faceplate that cannot be installed on any other type of device. 2.2. The device does not have mounting means to accept other configurations of faceplates. 2.3. The device is equipped with a nonmetallic yoke. 2.4. All parts of the device that are accessible after installation of the faceplate are manufactured of nonmetallic materials. 3. Connection to an equipment grounding conductor shall not be required for snap switches that have an integral nonmetallic enclosure complying with Section E3905.1.3 in the IRC code book.

You Should Know

- Faceplates provided for snap switches mounted in boxes and other enclosures shall be installed so as to completely cover the opening and, where the switch is flush mounted, seat against the finished surface.

SWITCHES AND RECEPTACLES

TABLE E4002.1.2 RECEPTACLE RATING FOR VARIOUS-SIZE MULTIOUTLET CIRCUITS

Circuit Rating (Amperes)	Receptacle Rating
15	15
20	15 or 20
30	30
40	40 or 50
50	50

© 2018 IRC®, International Code Council®

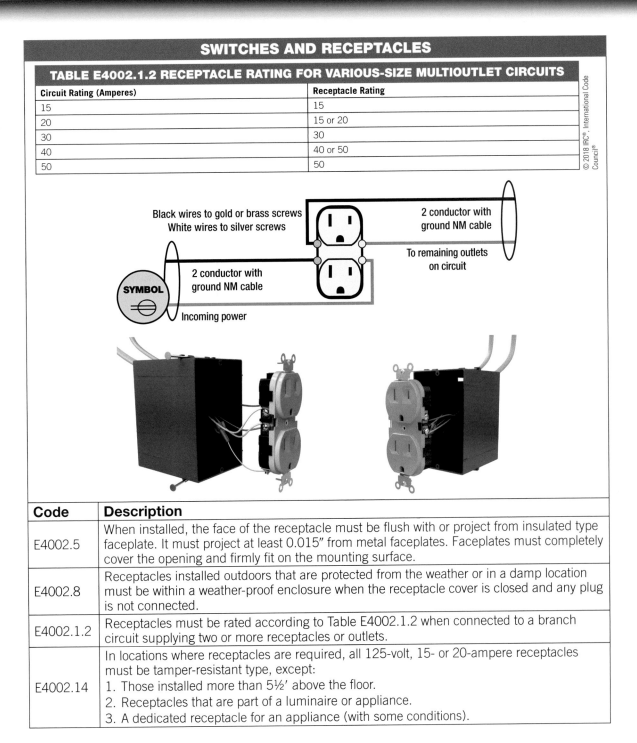

Black wires to gold or brass screws
White wires to silver screws

2 conductor with ground NM cable

2 conductor with ground NM cable

To remaining outlets on circuit

SYMBOL

Incoming power

Code	Description
E4002.5	When installed, the face of the receptacle must be flush with or project from insulated type faceplate. It must project at least 0.015" from metal faceplates. Faceplates must completely cover the opening and firmly fit on the mounting surface.
E4002.8	Receptacles installed outdoors that are protected from the weather or in a damp location must be within a weather-proof enclosure when the receptacle cover is closed and any plug is not connected.
E4002.1.2	Receptacles must be rated according to Table E4002.1.2 when connected to a branch circuit supplying two or more receptacles or outlets.
E4002.14	In locations where receptacles are required, all 125-volt, 15- or 20-ampere receptacles must be tamper-resistant type, except: 1. Those installed more than 5½' above the floor. 2. Receptacles that are part of a luminaire or appliance. 3. A dedicated receptacle for an appliance (with some conditions).

ELECTRICAL

SWITCHES AND RECEPTACLES *(cont.)*

Code	Description
E4002.15 and E4002.16	Receptacle assemblies for installation in countertop surfaces shall be listed for countertop applications. Receptacles shall not be installed in the faceup position in or on countertop surfaces or any work surfaces except where listed for the application.

You Should Know

- A receptacle shall not be installed within or directly over a bathtub or shower stall.
- There are three locations where tamper-resistant receptacles are not required: (1) receptacles that are more than 5.5' above the floor, (2) receptacles that are part of a luminaire or appliance, and (3) receptacles for appliances, cord-and-plug connected, located in dedicated spaces and are not easily moved.

LIGHTING FIXTURES

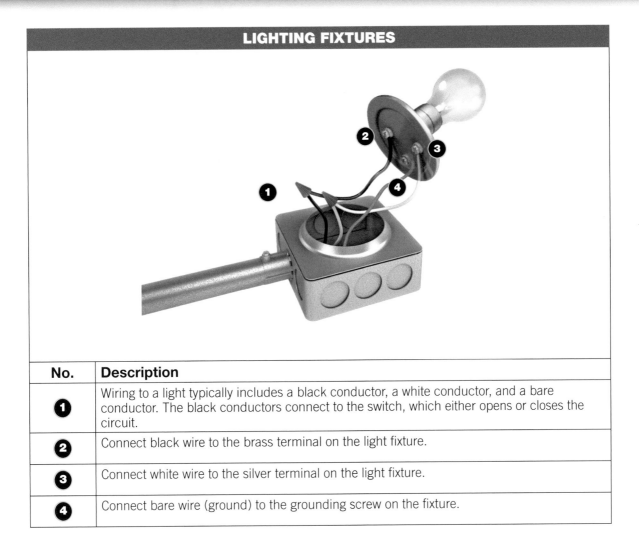

No.	Description
①	Wiring to a light typically includes a black conductor, a white conductor, and a bare conductor. The black conductors connect to the switch, which either opens or closes the circuit.
②	Connect black wire to the brass terminal on the light fixture.
③	Connect white wire to the silver terminal on the light fixture.
④	Connect bare wire (ground) to the grounding screw on the fixture.

ELECTRICAL

FIXTURES

The minimum clearance between luminaires installed in clothes closets and the nearest point of a storage area shall be as follows:

- Surface-mounted incandescent or LED luminaires with a completely enclosed light source shall be installed on the wall above the door or on the ceiling, provided that there is a minimum clearance of 12″ between the fixture and the nearest point of a storage space.
- Surface-mounted fluorescent luminaires shall be installed on the wall above the door or on the ceiling, provided that there is a minimum clearance of 6″.
- Recessed incandescent luminaires or LED luminaires with a completely enclosed light source shall be installed in the wall or on the ceiling provided that there is a minimum clearance of 6″.
- Recessed fluorescent luminaires shall be installed in the wall or on the ceiling provided that there is a minimum clearance of 6″ between the fixture and the nearest point of a storage space.
- Surface-mounted fluorescent or LED luminaires shall be permitted to be installed within the storage space where identified for this use.

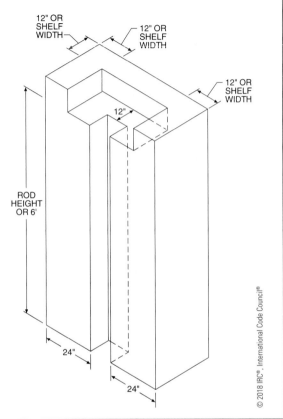

Figure 4003.12
Closet Storage Space
For SI: 1 inch = 25.4 mm, 1 foot = 304.8 mm

© 2018 IRC®, International Code Council®

Code	Description
E4003.11	Certain light fixtures are regulated for location within clothes closets or bath areas. Cord-connected luminaires, chain-, cable-, or cord-suspended-luminaires, lighting track, pendants, and ceiling-suspended paddle fans must not be within 3′ horizontally and 8′ vertically from the top of a bathtub rim or shower stall threshold.
	Lights within the actual outside dimension of the bathtub or shower to a height of 8′ vertically from the top of the bathtub rim or shower threshold must be approved type and marked for damp locations. Where subject to shower spray, these lights must be approved type and marked for wet locations.
E4003.12	Other lights installed in clothes closets are limited to surface-mounted or recessed incandescent lights with completely enclosed lamps, surface-mounted or recessed fluorescent lights, and surface-mounted fluorescent or LED lights identified as suitable for installation within the storage area.
	For a closet that permits access to both sides of a hanging rod, the storage space shall include the volume below the highest rod extending 12 on either side of the rod on a plane horizontal to the floor extending the entire length of the rod; see Figure E4003.12.

LUMINAIRE INSTALLATION AND TRACK LIGHTING

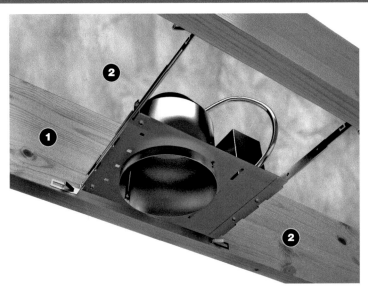

ELECTRICAL

LUMINAIRE INSTALLATION AND TRACK LIGHTING *(cont.)*

No.	Code	Description
1	E4004.8	A recessed luminaire that is not identified for contact with insulation (IC rated) must have all recessed parts spaced at least ½" from combustible materials.
		A recessed luminaire that is identified for contact with insulation (IC rated) is allowed to be in contact with combustible materials at recessed parts, points of support, or where it passes through a building's structure including finish trim parts at openings of ceiling and wall.
2	E4004.9	Insulation that retards heat cannot be installed above a recessed light fixture or within 3" of the lighting enclosure, wiring component, or ballast unless the fixture is IC rated.
3	E4005.4	Lighting track may not be installed in certain locations including: 1. Where it can be damaged. 2. In damp or wet locations. 3. Where it can be affected by corrosive vapors. 4. In rooms where batteries are stored. 5. In hazardous locations. 6. Where the track is concealed. 7. Where the track extends through walls or partitions. 8. Lower than 5" above the finish floor (unless protected from damage). 9. Where prohibited in bathrooms (Section E4003.11 in the IRC code book).

FLEXIBLE CORDS

Flexible Cord Lengths

Appliance	Minimum Length (Inches)	Maximum Length (Inches)
Kitchen Waste Disposal	18	36
Built-In Dishwasher	36	78
Trash Compacter	36	48
Range Hoods	18	48

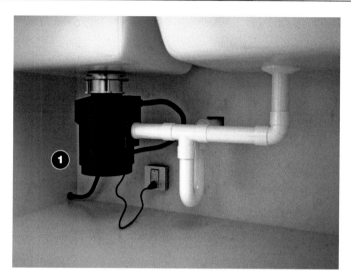

No.	Code	Description
1	E4101.3	Flexible cords and plugs may only be used where the appliance is listed to have the connection. The cord must be identified as suitable for such use in the instructions provided by the manufacturer.
	E4101.5	All appliances must have a means of disconnection of all ungrounded supply conductors.
		All switches and circuit breakers used as a means of disconnect must have an indication on the switch specifying on or off. The means of disconnect must meet the requirements of Table E4101.5 in the IRC code book.
	E4209.4	Electrical equipment on hydromassage tubs must be accessible for service or replacement. The access provided must not damage the building structure or finish. Where there is a cord and plug connection for the hydromassage tub, the face of the receptacle must be within 12″ of the access opening.

You Should Know

- Appliances and equipment shall be installed in accordance with the manufacturer's installation instructions.

ELECTRICAL

OHM'S LAWS

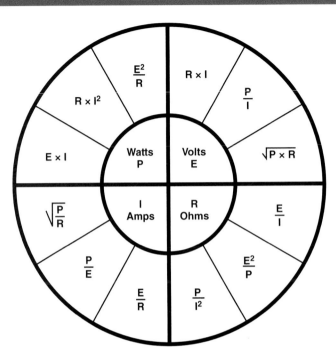

P = Power measured in watts
I = Current measured in amps
E = Electromotive force (EMF) measured in volts
R = Resistance measured in ohms

Example: A clothes iron is rated at 1400 watts and connected to a 120-volt
circuit. How much current does the iron draw?
I 5 P/E 5 $^{1400}/_{120}$ = 11.66 amps
Example: What is the resistance of a clothing iron?
R 5 E²/P 5 $^{120²}/_{1400}$ = 10.3 ohms

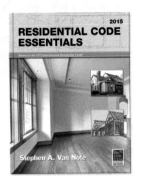
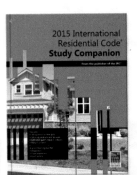
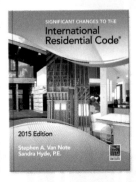